Mathilda Barnett

The new biology

The true science of life

Mathilda Barnett

The new biology
The true science of life

ISBN/EAN: 9783337216894

Printed in Europe, USA, Canada, Australia, Japan

Cover: Foto ©berggeist007 / pixelio.de

More available books at **www.hansebooks.com**

THE

NEW BIOLOGY

— OR —

THE TRUE SCIENCE OF LIFE

— BY —

M. J. BARNETT,

AUTHOR OF "PRACTICAL METAPHYSICS, OR THE TRUE METHOD OF
HEALING," "HEALTH FOR TEACHERS," "JUSTICE
A HEALING POWER." &c.

BOSTON:
H. H. CARTER & KARRICK,
No. 3 BEACON STREET.
1888.

CONTENTS.

PREFACE.

WHATEVER is new is also old. We speak of new foliage on the tree in spring-time, although we are well aware that the germ, the soul of that foliage is always alive within the tree, and that it only manifests itself at certain seasons. We know that during the season of rest it is just as much alive as during the season of activity, and that it is as old as the tree itself.

All truth in man is coeval with man, and only outwardly manifests itself at certain due and divinely appointed seasons. Were it not already within it could not become manifest without.

Although the truth is always within us as the germ of the foliage is within the tree, and will like that foliage become manifest at certain divinely appointed seasons, yet, as it is for us to tend and nourish the tree that the foliage may be more flourishing and abundant, so it is for us to foster and develop the truth that is in us, that it may yield us richer results.

The season for manifesting the truth that is in us may have already arrived, but if we do not co-operate with divine intention, and work in favor of this manifestation, our foliage, though it may appear, will be sparse and sickly.

It does not matter if this foliage reappears with certain changes of form and color. Truth may be all the more acceptable for its varying manifestations, and may thereby attract more attention from those who have become almost unconscious of its existence.

If we give an old truth a new dress, we do so, only as a courteous plucking of the robes of the passer-by. We would not mislead, but would only awaken the mind to a consciousness of ever-present truth, which from its very familiarity has ceased to make any impression.

Let us go back in imagination to the days of Saxon-English. A mother cries out to her children to come to the window and see an omnibus. "An omnibus!" they echo, trying for the first time to mouth the new Latin word, and leaving their play to see what manner of thing it is. "What big wheels it has!" they cry. "How strong it is!" "How many people it holds!" But suddenly recognizing the familiar object, they say, perhaps with a shade of contempt, "Why it is only a carryall!" But then, I never noticed that it had such great springs, and such handsome red wheels, and such a nice top for trunks and boxes." So they continue to note its peculiarities and descant upon them, and all because it has received a new name.

Such children are we. We need rousing up to a new view of old familiar truth.

CHAPTER I.

THE OLD AND THE NEW.

In the old biology of our day, great attention is paid to all the external manifestations of life as exhibited in motion and force. The more material of the *savants* will say that matter generates its own life, while those whose spiritual perceptions are more awakened will feel that all life manifested through matter, proceeds from spirit. But very few of either class seem to realize that from whatever source this life force proceeds, they, themselves are its masters. They fail to regard themselves as the engineers of this great motive power.

In the new biology, or the true science of life, which is to-day being presented to us under so many aspects, we are given to understand that we are in charge over this motive power as manifested

(9)

both in ourselves and in the world around us. We are the engineers whose duty it is to see to it that a due supply of force is obtained and rightly directed so that the working of life's machinery may accomplish its preordained end.

What would be thought of an engineer who apologized for the feeble and inefficient working of his machinery, by saying that his steam gave out?

Would he not be told that it was for him to see to it that he had a due supply of steam and also that it was rightly directed instead of being allowed to escape to no purpose?

If we lack the life force, which is ever at our command, the lack is in ourselves, and should not be attributed to anything outside of ourselves.

In the true science of life, we must look through and beyond the outward manifestation of life as visible in the material body, to the working of spirit within, to that workshop of all that is visible in us externally. We must not be startled at what appears to us a new way of regarding things. We must not feel that our old way of thinking is necessarily the best way, for a new way is quite likely to be an advance on an old way.

Lot's wife looked back and turned into a pillar of salt. Turning away from the new and looking back upon the old is a petrifying process, and its effect upon us is well symbolized by a pillar of salt.

Throwing a false glamor upon the past and underrating the present, is a stumbling block in the way of progress. It deters one from fully appreciating and utilizing the present.

That expression often uttered with a sigh, "The good old days," casts a reflection upon the present. It implies that yesterdays are better than to-days, which is a great mistake. To-day is better than yesterday, and to-morrow will be better than to-day. In ascending the spiral of inevitable progress from age to age, we gain a view of what is best by looking forward and upward, instead of backward and downward.

The world and its inhabitants are further advanced to-day than ever they have been before within historic ages. There are always certain individuals who are remarkably in advance of their race. There have also been ages in the past, in which certain races have been remarkably developed in some one direction. They have perhaps been far beyond us in certain arts and sciences, but as a whole they have not been so developed, so near a perfect comprehension of the truth and intention of being as we in this later day.

Some old nations possessed an almost perfected knowledge of material things, which has seemed to die with them, but have we reason to believe that they had passed through their material development and stepped up on to a spiritual plane above our present one?

The ancient Pompeians may have possessed the secret of exquisite and durable colors, but were they able to color their lives with that spirituality which would give the very highest prosperity?

The Grecians of old may have reached a high intellectual culture, they may have excelled us in the plastic and other arts, but did they possess that knowledge of spiritual things, which alone ensures continued prosperity? Have we any reason to suppose that they had as a whole passed our point of progress?

One, as a child of ten years may be able to spin a top or fly a kite more dexterously than as a man of forty, but would we consider that the individual had therefore retrograded instead of advanced?

We are all on our way through matter to pure spirit, and, like the earth in its diurnal journey, we are not at any one time flooded with light upon our whole being, but receive it upon one little part after another until our material day is consummated.

Further back than any people of which we feel that we possess accurate knowledge, we can imagine races much more spiritual than we; but if they had not passed through a certain material experience, were they more advanced than we? An infant may be more innocent than an adult, but is he therefore more advanced? Is not innocence more valuable when it is coupled with knowledge? Do we consider a human being, however innocent he may be, further advanced with one year of this life than with fifty?

Why are we prone to think the past better than the present? We forget the evil and remember only the good of the past. We forget the flaws and remember only the beauties. In accordance with a benificent law of our being, having no use for evil, it drops away from us as we pass along.

Whether we remember it or not the good of the

past is all with us, and we are increasing its store by our daily addition to it.

Why should any one imagine that in a new dispensation, we are deprived of some of the good of the old dispensation? In a new illumination of our minds with spiritual truth, the old good is only supplemented with a new good that is as much better and as much richer as we are able to receive.

Why should it be thought that we, in the present day, are deprived of certain spiritual advantages that were enjoyed in ages gone by? We are never deprived of any spiritual good that has once been given us. What is once ours as property of the soul, is ours forever, and ours to increase and not decrease. If we let it remain latent within us, to our own undoing, instead of being bereaved we are only guilty.

It is frequently said that eighteen hundred years ago there was bestowed upon the world a knowledge of the power that could work miracles in the way of moral and physical reformation; but that it was a power peculiarly the property of that age and not intended for us at the present day.

We are not able to see that Jesus, either by precept or example, ever intimated anything of the kind. He taught his disciples how to do marvelous work and told them that they should do greater work than he did. We think that his promises to twelve of his disciples are intended for all of his disciples throughout all time, and are not all who follow his teachings his disciples? Our work may not be greater in kind, but it can be greater in extent. If a man gives a certain kind of knowledge to a whole country, he is said to be doing a greater work than if he gave it only to a few individuals. The truth that Jesus shed upon a few we can shed upon many. If his followers of eighteen hundred years ago could so harmonize themselves with divine law as to become a healing power to others, why should not we in this age, be able to do likewise? Are we to suffer loss from the fact of coming into the world at a more advanced period of its course? It does not seem rational to suppose so.

The distinctions made by Jesus in adopting the law were never between the new and the old, but between the true and the false. He adhered to the portions of the old law that were in accordance with truth, and rejected those that were false. He never revered the law because it was old, and he never set it aside because it was new. He sifted the old law and used the wheat and threw away the chaff.

He endeavored to replace the old Jewish conception of a God capable of jealousy and anger, for the higher conception of a Divine Parent all love and wisdom. He accepted the decalogue, old though it was, but endeavored to give it a richer and fuller meaning. He taught that not only he who killed, but he who was angry with his brother, should be in danger of the judgment; that instead of performing unto the Lord our oaths we are not to swear at all; that instead of an eye for an eye and a tooth for a tooth, we are to resist not evil; that instead of loving our neighbor and hating our enemy, we are to love even our enemy. Wherein the old law was false, he contradicted it; wherein it was limited he extended it; wherein it was just and true he enjoined it upon his disciples, at the same time shedding upon it such light as disclosed to their vision a more interior and spiritual meaning within the too-revered letter of its truth.

All the teaching that Jesus, during his sojourn on earth, gave to the world, belongs to us to-day.

If we do not seem to have it, it is because we have mislaid it. If we do not know how to apply it to our lives and make it practical, it is because our powers in that direction have become crippled with disuse. The teaching of Jesus, being the pure essence of truth, was intended for all time. When we are capable of receiving an addition to that teaching, it will be given us.

There is no reason why the marvelous works of moral and physical reformation that have been performed in the past, should not with the same conditions, be performed by us to-day. There is no reason why the same conditions should not be commanded by us as perfectly as they have been commanded by others in the past.

We have conclusive proof of our powers to-day in the demonstrations that are constantly taking place all around us. Moreover we are each one of us capable of furnishing demonstrations for ourselves and in ourselves. Whatever powers for good we are in possession of, are powers intended for us to use. Whatever good we can accomplish by these powers, is a good intended to be accomplished. God created none of our faculties in vain.

It is a law in all progress, both spiritual and material, that we lose nothing good on our way upward, but only add new good to the old.

In what is usually termed nature, that is God manifest in matter, we see that the vegetable kingdom possesses all that is contained in the mineral kingdom with new developments of its own; and still further that any one vegetable contains many minerals. So the animal kingdom contains all of the vegetable kingdom, with new developments of its own; and still further, any one animal contains a combination of vegetable elements. When we behold that high spiritual manifestation called the human mind, we find that it contains all there is in the material world with vastly more in addition. Every mineral, every vegetable, and every animal is in the mind of man, and not only in man as a whole, but in each individual man. In the mind of every human being is also the mind or instinct of every animal. Each animal is endowed with one characteristic mental trait; but in man, are, either latent or developed, all mental traits. He has within him the courage of the lion, the meekness of the lamb, the fidelity of the dog, the industry of the ant, and the ingenuity of the bee.

Nothing good is ever dropped out or lost on the way, in the universal grand march of progression, and every spiritual entity as well as every material manifestation has set out upon this march, and in spite of their little temporary backslidings and their seasons of rest, they are ever going forward and upward.

The new good, that we pick up along the way, is only an addition to the old, which is always ours, with just as much more as we are able to carry.

CHAPTER II.

To-morrow is higher up on the spiral of advancement than to-day. Although to-morrow may be more full of opportunities than to-day, still, *our* to-morrow depends upon the use *we* make of to-day.

If a child educated in an institution that has risen from an a–b–c school to an incorporated and chartered college, fails to learn therein anything more than his alphabet, he reaps no benefit from the advance of the institution. It is only his individual work that can benefit *him*.

If we, in a more spiritual age of the world, fail to rise up out of a material condition, we receive no benefit from the world's advance. If we would

(20)

be advanced in the future, we must advance ourselves in the present.

To-day we sow for to-morrow's reaping. The
present is the all important time. It is the only
time that is within our grasp. It is the only time
that is at our disposal.

It seems a grave error in any system of philosophy or religion, to lament this mortal life which is
now ours, to regard it only as an affliction to be
endured, with resignation, and to feel that all happiness lies in the future. Our future happiness
depends upon our present condition of harmony
with divine law.

There is much cant in the various religious sects
of the day to the mistaken purport that the sooner
we are taken out of this life the better it is for us,
as though God had made a mistake in placing us
here. Thousands among the ignorant are encour-
aged in this morbid sentimentality, by the fervid
hymns that extol the by and by as though it were a
blessed escape from the inevitable miseries of the
present life.

It is a curious fact that it is principally among
devout Christians that we find this mistaken view
of life. It is they who seem to think that the

sooner God repairs his blunder in placing us here, the better it will be for us. We fail to discover any such view, in the teachings of their professed master and Christ. They seem to think that all unhappiness resides in their bodies. We even know of a certain theologian of note, who asserts that sin is of the body, and when we drop the body we shall come into purity and happiness.

Sin no more resides in the body than it resides in the clothing, and one is free from sin no more because he has cast off his body, than because he has cast off his clothing.

One may sin in order to pamper his body, as he may sin to decorate his clothing, but the sin is in his spirit, his mind.

Suppose, for illustration, that a father so mistakenly loved his son that he committed crime in order to lavish luxuries upon him, but that when the son died he ceased his criminal practices because there was no longer a motive for them. Should you consider that this man became suddenly virtuous because he lost his child? Without a change in his interior condition, would he not, with sufficient incentive, be ready to return to his crimes?

The wrong, the evil was in the father's mind,

and remained there, whether quiescent or active, until overcome and driven forth.

Now, our body is our child, whom we so mistakenly love, that we are ready to commit wrong for its sake. But the wrong, the evil is in our own spirit and mind, which we carry with us when we leave our body behind.

"As the tree falls so shall it lie." We, however, do not suppose that as the tree falls so shall it lie *forever*. The condition in which we leave this mortal life shall be our condition on our arrival in the spiritual world, though we do not feel that we need remain in that condition *forever*. But is it not rational to suppose that we will remain in that condition until we desire to rise up out of it into something better?

Casting off the body, like casting off the clothing may give us more freedom of action, but it cannot at once work a change in our interior condition. Without the body precisely as with the body, we can reap only the fruits of our spiritual condition.

There is just as much loving wisdom in that part of God's plan which places us in this material world as in that part which takes us out of this world into the kingdom of pure spirit. The world's latest and greatest teacher has told us that the kingdom of heaven is within us. We can live in that kingdom now without waiting to drop our bodies. If we were living wholly in that kingdom we would be so in harmony with our material surroundings that we would have no desire to sentimentalize over getting rid of them. We would be able to perceive that this life was intended for our happy school days to prepare us for something higher. Our bodies reflecting the harmony of our minds would be for us the perfect tools they were intended to be, and so beautifully adapted for our use that they would afford us only pleasure.

If our bodies are a burden to us, there is something wrong with that higher part of us, which has charge of these bodies. Instead of lamenting that our body is a hindrance to us, let us lament and overcome our errors and then our bodies will not obtrude themselves unpleasantly upon our notice.

The material body of man is no more to be despised than any other material thing. On the contrary, it is the most wonderful, the noblest of God's material manifestations. When it was created it was good, like all God's works, and if we have misused it and it has become bad, the fault is ours.

While we speak of some who underrate the things of this material life, we know there are many more who so overrate them, that they think this life is the all in all.

It is astonishing to see a person of perhaps advanced age, manifest no interest whatever in the immortal part of his nature. He may fully believe that in less than ten years he will surely be called to live in a spiritual realm, a kingdom in which material things have no place, yet his whole interest continues to be absorbed in these material things. The things of the spirit, which will so soon be *forced* upon him, fail to seem of any importance to him.

What would be thought of a business man shortly to be established in a foreign city, who took no trouble to learn anything of the language or customs of the people among whom he was soon to dwell, and who, if he chanced to be where valuable information concerning them was being imparted, would, instead of listening, turn indifferently away.

His attitude would doubtless be considered an unaccountably foolish one. Yet how much more foolish is the attitude of a person, who permits himself to enter into the spiritual world, a stranger in a strange land, knowing nothing of even the alphabet of the new language by which alone he can hold communion with his fellow beings.

As soon as our thoughts aud interests are upon spiritual things, we are living in the spiritual world. When we begin to unfold the spiritual part of us, when we begin to develop this only real, lasting part of us, we are learning the language of the country to which we are going, and if at any moment we are called upon to embark for it, we may feel sure that we shall there find ourselves at home.

How much better and wiser it is for us to enter into the spiritual world *now*, instead of waiting until we are forced into it by dropping our material bodies, or, what would better express it, instead of waiting to pass into a realm, where all is spirit, while our minds can conceive of only matter, and we are therefore left without anything.

The folly of not being interested in spiritual things while in this life, would seem sufficiently great if the future alone were involved. But when we learn that one who refuses to enter into the kingdom of heaven that is within him, while he still has his material body, deprives himself also of present happiness and injures the very body he adores, we feel that his folly approaches idiocy.

Even if a man loves only material things, it is but the part of wisdom so to rule his spirit and mind that these material things may afford him the utmost enjoyment. " Blessed are the meek for they shall inherit the earth." It is only those who maintain a lowly and correct attitude of mind who can command the fullest enjoyment of the things of earth.

Ruling the lower nature with the higher to gain material enjoyment is certainly not working from the highest motive; but the practice of good begets a love of good. When we once love good we will seek it for its own sake, and our happiness will reside in the good rather than in any minor benefit it confers upon us.

For illustration, we can easily understand that a man in an inverted condition of mind may, in beginning to seek money, desire it only for the happiness he falsely supposes it will bring him; but in his eager pursuit he shortly loses sight of the end, and finally loves money for itself alone.

Now, the same law that works in our downward inclination, works also in our upward growth. When we pursue good with even an oblique motive, we will end in loving good for itself, which is the true motive.

When we look within ourselves for our spiritual nature, even though we may do so only for the repairing of our bodies, we shall be likely to catch at least a glimpse of it.   As soon as we desire to unfold that real part of our being, we shall be so aided in our endeavor, that we shall be enabled to succeed.

A sincere desire for any good is always an energy set to work to accomplish that good.   It is a cry to all the powers of good for aid, to which there is always a helpful response.   Let us have this desire, let us utter this cry at once.   *Now* is the day of Salvation.   We are not to work with a view to saving ourselves in the hereafter, but with a view to saving ourselves in the present, or rather with no view to self at all, but with only a view to harmony with divine purpose, and the result will be immediate salvation.

It is to the evils that are holding us in bondage to-day that our efforts are to be directed. The future will take care of itself.

We need not trouble ourselves about what is going to be our condition when we cast off our bodies, for our condition will then be just what we are making it now while we have possession of our bodies.

There is no death for anything except evil, and the evil that is in us does not die until we kill it.

If you morbidly imagine that the evils in your mind, which are so reflected in your material body as to make it a burden to you, will suddenly drop away from you with that body, you will, we fear, be wofully disappointed on reaching your future home. Your evils are no more destroyed with the disintegration of your material body, than your body is destroyed with the breaking of a mirror that reflects it.

Even when the harvest time comes we cannot reap what we have not sown.

Instead of underrating this material life, let us feel that it is a perfect instrument, wisely and lovingly adapted for our use and enjoyment. Let us also thank God that it is so sure an indication of our interior condition, that however spiritually-blind we may be, we yet can always, with our material eyes, behold in these bodies the external manifestation of our every inward error, and can thereby become enabled to free ourselves from bondage.

This material life should be one grand unbroken harmony, even though at times in a minor key.

This material body should be a musical instrument perfectly attuned to the harmonies of this life.

CHAPTER III.

Spiritual science enjoins upon us the necessity of harmonizing ourselves with the law. To what law does it refer? It refers to the universal law of God, the divinely established order that governs the universe, the fixed relation between cause and sequence.

We speak of God's law as one law, in the same way that we speak of civil law as one, though we well know that divine law, like civil law, is an aggregation of many laws.

The term inexorable law, has to many of us a harsh sound. We fail to discover tenderness in a Divine Parent, who does not modify his law in response to our entreaties and our supposed necessities. This limited and ignorant view of divine

(31)

order .leads us to feel that we are so restricted that we have no free will. But within the circle of necessity, the great circle of God's law, is always the smaller circle of man's free will, and within that smaller circle is all the freedom that man is capable of enjoying.

If we are wise and upright in spirit, we will not struggle against the law; we will so harmonize ourselves with it, that, instead of finding it bondage, we shall discover it to be the utmost freedom.

Which man enjoys the most freedom, he who pursues an unlawful course and devotes his best energies to dodging the civil law, whose penalties he must finally experience, or he who devotes his energies wholly to good and thereby becomes so harmonious that he is scarcely conscious of the existence of law? The former course is slavery; the latter, freedom.

The laws that govern the material world are called physical laws, or natural laws; but the laws that govern pure spirit are just as natural, just as much born of God, as the laws that govern matter. In reality all law is spiritual, whether it governs matter or pure spirit. It is just as natural to love our fellow beings, as it is to inhale air into our lungs, yet one is a spiritual, and the other is a physical act. But the power that makes the physical act possible, is a spiritual power. All adjustment of cause and sequence is an outcome of spirit, of mind. Even civil law is an outcome of the minds of those who ordain it.

We none of us think of contradicting or struggling against so-called physical law; we harmonize ourselves with it. We do not try to prevent the wind from blowing, but we spread open the sails of our ships that it may aid us in our commerce, and when it rises into a tempest we study to avoid disaster by protecting ourselves against it. We do not struggle against the law of the solar system that governs the order of day and night; we simply adjust ourselves to it.

What would be the prosperity of a farmer who struggled against the inevitable course of nature in the order of the seasons, who determinedly planted his seed in the autumn and stubbornly expected to reap his harvest in the spring-time?

It is just as great folly to resist spiritual law as it would be to resist physical law. The law which ordains that suffering and disease shall follow upon sin and error, is just as fixed, just as inexorable as

the law which ordains that darkness shall follow
upon the turning away of the earth from the sun.

The one law we are able to see clearly; the other
we do not perceive. Why is this? It is because
our spiritual perception, which is required in order
to see spiritual things, is not yet unfolded.

Spiritual science is a knowledge of law as governing in the realm of spirit, and through spirit
descending into the realm of matter.

True spiritual science is never coercion. It is
simply loving conviction. It explains the working
of divine law, but never enforces arbitrary personal law. It never *forces* away error. It never
commands us to drop beliefs, however false they
may be, for as soon as we come into knowledge
our false beliefs, our erroneous creeds, and our
mistaken views of the science of life, become dis-
sapated as naturally and as easily as the shadows
of a dimly-lighted room disappear upon the intro-
duction of light. Before we have the light in the
room we confound the shadows with the objects;
so when we are benighted, we confound error with
truth, and then we hold fast to the error. True
science will not strive to take away our creeds, for
so long as they comfort us and we hold fast to

them, we need them however erroneous they may be. They are sure to have *some* good in them.

If you take away a lame man's crutch before he has anything to replace it, he will fall. He needs his crutch until he finds something greater than a crutch. True science will offer him that something greater, but will not expect him to take it in the place of his crutch, until he sees for himself that it *is* greater, and when he does see this, he will throw it aside without a pang. Our beliefs, false though they may be, cannot be *wrenched* away from us; but when we rise above their level, we simply come up out of them, and leave them behind without a pang.

True spiritual science does not attempt to enforce any human law; it merely teaches the working of divine law. It teaches us to have divine patience with ignorance and error, for we have, none of us as yet, come into all knowledge. It takes no thought of nationality or of sect, but is so broad that it opens its doors to every spiritual being.

The teachings of this science are intended so to help us unfold our spiritual powers, that, instead of accepting its truths blindly, we are able to discover them for ourselves, and no truth is ever ours until we have discovered it for ourselves. It is by our own spiritual apprehension that we must seize upon truth. It is by our own spiritual powers that we must assimilate truth. The workings of the law can be pointed out to us, but we cannot fully accept them as truths until we have, to some degree, demonstrated them upon ourselves and others.

A blind man may believe there is such a thing as color, but color is not real to him until he can see it for himself. Spiritual science works upon the spiritual vision in order that we may perceive its truths by making them real to ourselves. Spiritual science opens the eyes of the blind.

We cannot change the law, and could we see clearly the beneficent intention in the very penalty of error, we would not even wish to change it. We would simply desire to change ourselves in our relation to it, and then our error would be changed to right thinking, the result of which would be happiness instead of pain.

A citizen does not feel that he is in a safe and protected condition until he knows something of the laws of the country in which he lives, and when he learns the laws he feels the necessity of harmonizing himself with them. If he learns that his residence within city limits must be built of brick or stone, he does not every year attempt to put up a wooden structure, and every year complainingly see it razed to the ground by order of the city authorities. He simply conforms to the existing law, which he cannot alter, and which, with sufficient insight into its intention, he discovers to be a beneficent law.

Why do we not pursue the same rational course in regard to divine law? We fail to realize the importance of placing ourselves in a safe and protected spiritual condition. We fail to realize that we have anything to do with our spiritual condition. We seem to consider ourselves in the hands of either a blind fate or an arbitrary ruler with whom it is useless for us to contend. We do not need to contend with the law or with the divine ordainer of the law, or with anything but the evil and ignorance within ourselves.

While we cannot alter the law, we can and must learn the law. There are two ways of learning the law; one through suffering and disease born of sin and error; and the other, through that peace which follows observance of the law, overcoming the world, rising up out of the bondage of material things. We can learn the law by defying it, or by complying with it. Which of these two ways shall we choose? It is only a mistaken and blinding love of self that leads us to choose the hard way, for the way of the transgressor *is* hard. Still, those who will not, or as yet cannot come into knowledge through harmony, must come into it through discord. The time always comes when discords resolve themselves into harmonies. There are always some in our midst who must have their discords, but it is our work to resolve these discords into harmonies for them.

Jesus tells us that the poor we always have with us. We shall always have with us those who are spiritually poor, as well as those who are materially poor. Poverty is a comparative term. We are all of us comparatively poor in spiritual possessions. While there are many at this moment, who are poorer than we, there are also countless beings in higher spheres, who have infinitely more of these spiritual possessions than the richest of us. So long as God's children are in different stages of their unfoldment, just so long there will be those of the same generation, of the same race, and even of the same family, who are rich, and those who are poor in spiritual things. But let us, instead of glorying in our own acquisitions, only endeavor to feed the hungry and clothe the naked. Let us shed the same love and tenderness on those who are behind us upon the road, as on those who are in advance of us. Let us follow the example of our Master and teach the law in all love and charity. Let us unceasingly pass on to others the little we may have learned of this divine jurisprudence. Let us feel that while suffering and disease are a necessity to those who are in a condition to have them, yet it is our bounden duty to endeavor to abolish that necessity by bringing them up into a better condition. Let us by teaching the loving purpose of the law bring about that harmony which offers the only freedom from suffering and disease.

CHAPTER IV.

CORRESPONDENCES IN DISEASES.

The material world is created by the spiritual world, and corresponds to it, as the shadow corresponds to the object that casts the shadow.

Sweedenborg says that the whole natural world corresponds to the spiritual world, and not only the natural world in general, but also in every particular.

The interior part of a man constitutes his spiritual world, and it corresponds to his exterior or his material world.

This truth forms the basis of all spiritual healing, which is the healing of the material body by first healing the spiritual part of us, which dominates that body, and to which the condition of that body must inevitably correspond.

(40)

Nations that are aggressive and warlike in spirit, will devote their best energies to the invention and construction of implements of war, and they will furnish and surround themselves with these implements. Their material surroundings will correspond to their spiritual condition. A simple people with a peaceful and home-loving spirit will engage in peaceful and quiet occupations, such as agriculture and the mechanic arts. A nation that is more aspiring in mind will engage in the fine arts; while one still more developed intellectually and spiritually will find their highest pleasure in the study of philosophy, ethics and religion, and the practice of good.

As it is of nations, so it is of individuals, whatever love or desire holds possession of the mind, gives form and color to the material condition, which is only a pushing out of the mind into externals.

These externals are sometimes said to be of no importance, and in many cases are denied an existence. Although they are the most outward effect of the working of the mind, the grossest and most transient of manifestations yet they have their uses.

They are an index of the controlling mind or minds that produce them.

We do not despise a thermometer because it cannot produce heat or cold ; but we value it as an index of temperature.

In the science of spiritual healing, the condition of the material body is important to us as an index of the mind or minds that brought about that condition.

The material body corresponds, not only in a general way to its dominating mind, but as Sweedenborg says of the material world in its relation to the spiritual world, it corresponds also in every particular, whether we are always able to trace this correspondence or not.

It is said in material science that each organ of the brain dominates a corresponding part of the body. But as the brain is itself only a material part, it cannot *dominate* any other material part. We would say that certain attributes of mind, working through its direct and most sensitive medium, the brain, act upon certain corresponding parts of the rest of the physical organism.

We all know that the affections act upon the heart, and to say that one who is sorrowful is heavy hearted, is not a mere figure of speech, but a literal truth, for his heart, at such times, performs its functions in a slow and labored manner. One who is joyful, is said, with equal truth, to have a light heart, a heart that briskly performs its functions

It has been noticed that the intellect acts upon the lungs.

The memory seems to be connected with the digestive organs. When the mind receives a great deal that it cannot digest or assimilate, and therefore cannot retain, the digestive organs will be quite likely to be taxed by food that they cannot utilize.

When the mind is so inharmonious that it cannot throw off harmful or useless conditions, when it clings to its errors, the excretory organs will be found unable to perform their functions in throwing off harmful or useless material from the physical body.

The organ of hope, for example, is said to preside over the liver. Melancholy, which is inverted hope, causes the liver to be slothful in the performance of its duty.

Caution is said to preside over the ear. An acuteness to perceive coming danger and to provide against it, will manifest itself physically in acute hearing. The hare, the most cautious of animals, is also the most acute of hearing. He is not acute of hearing, because—as is sometimes said—of the favorable construction of his ear, but, on the contrary, his attribute of caution has created the peculiar construction of his ear.

Not only man as a collective whole has his duty in life, but each man as a separate being has his special duty marked out for him. In like manner, not only the mind as a whole has its duty in relation to the physical body, but each separate attribute of mind, has its special duty marked out. As one man cannot fail in his duty without affecting the whole brotherhood of man, so one attribute of mind cannot fail in its duty, without affecting the mind as a whole, and, through that mind affecting the physical body as a whole. It is this general demoralization, this sympathetic response of one part to the cry of another part, that renders it so difficult for us to trace the working of the law of correspondences.

In searching, however, ior the special cause of a special ailment, we must go back through generals to particulars.

A particular ailment in the physical body, will, when rightly traced, be found to correspond to the particular cause in mind that created that ailment.

You may think that as your ailment is hereditary, and was created by one of your ancestors, and does not correspond to any of your mental conditions, that you are not responsible for it. You are not, in this case, responsible for its creation, but you are responsible for its perpetuation. If it has been brought to you, it is your work to reject it. If it has been put upon you, it is for you to destroy it.

If you bought a defective piece of goods, would you consider that the merchant of whom you bought it, offered a legitimate excuse for selling the bad work, in saying that he did not make it? No. You would feel that while he might not be responsible for the making of it, he certainly was responsible for receiving it, and for selling it, and that in doing so he was injuring himself and all with whom he had dealings. Just so it is with us concerning all bad work classed under the head of disease. We are fully responsible, not only for the ailments that we create ourselves, but also for those of another's creation, which we harbor to the injury of ourselves and of all others who come under our influence.

In referring the law of correspondences to the creation of disease by mental conditions, spiritual science does not hold you responsible for the *creation* of any disease that is put upon you by another; and the correspondence is between the ailment and the creating cause, and not between the ailment and secondary cause in you, which is your ignorance or weakness in harboring the ailment.

For illustration, your mother created consumption in herself, by yielding to a consuming grief. The grief in her mind produced a corresponding inharmony in her body. Now, your inherited tendency in that direction is fostered, not by grief at all, but only by ignorance of the possibility of ridding yourself of it, and by your constant fear of it, or the fear of those around you.

It is well known in medical practice, that those who indulge in impure thoughts—which lead to impure acts—will reflect a corresponding impurity in their bodies in some eruptive and loathsome disease. It is also known that they who indulge in inflammatory thoughts, create inflammatory diseases in their bodies.

But there are a thousand and one ailments traceable to their corresponding causes in mind, whose workings are so subtle, and so much in the realm of the unseen, that these causes are never discovered or even suspected by the practitioners, who direct their attention chiefly to externals.

Sometimes a medical doctor is astonished and confounded at the persistence of an ailment, for whose existence, to say nothing of its continuance, there seems to be no cause. His research does not extend to the realm of causation, and the correspondence of the ailment to its cause in mind is perhaps so complex and subtle that only an *adept* in spiritual science could trace its working.

If a man in a fit of rage bursts a blood vessel, even the most material doctor would trace the physical result back to a cause in mind, because the working of the law would be so immediate and so apparent. If a man should fall dead from sudden fright or sudden joy—of which there are numberless cases on record—every one would be ready to admit that he had been killed by a thought.

For illustration, a poor old Italian woman, who had invested her last few pence in the Royal Lottery, when suddenly informed that she had won a fortune, fell dead at the feet of the messenger. The me enger did nothing to her body. The thought he transferred to her mind in the few words, —"Your numbers have won a thousand francs," —was too sudden and strong for her mind to bear, and the shock to her mind instantly extended to her physical body, and death was the result. Even the ignorant peasants that surrounded her were well aware that she had received her death blow through her mind.

If we form the habit of tracing a bodily ailment back to the mind, whether we at once see any connection or not, we shall be able to discover, not only that it was created by a condition of mind, but also by a condition corresponding to the ailment.

A young girl was suffering with that distressing malady called asthma. It was known that it was not hereditary with her, and as the girl had comfortable means, and, what is considered an easy life, and otherwise good health, there seemed to be no reason for the ailment. A change of climate gave her temporary relief, but of course effected no radical cure. How did she create such an ailment in herself? The healer, who took charge of her case, felt sure that there must be a corresponding condition of mind back of it, as a cause. She soon learned that the young girl, though living in the midst of agreeable people, even of her own age, was nevertheless very lonely, that her highest thoughts and warmest feelings were repressed. The best part of her nature was shut in and stifled, just as her lungs were stifled in breathing. The girl did not realize her mental condition until it was set before her in the new light of spiritual science, and then she saw no remedy for it. She was without a family tie and had no congenial friend.

The healer of course ignored her difficulty in breathing. For it had indicated the real malady and was of no further importance. Like a true soul physician she urged her to open her heart and her mind, and her breathing would take care of itself. She told her that there were thousands of her fellow beings, who were starving for just such good thoughts as she was repressing in her mind, and for just such rich affection as she was shutting up within her heart. She advised her not to wait for them to come to her, but to go

forth and seek them and not rest until she found
them for they would be her remedy. She told
her to begin without delay to utter aloud, even in
the solitude of her room to the empty air around
her her strongest thoughts and feelings, so as to
give them forth, to express them outwardly, and
they would be sure to reach some one and do
good, which good would at ouce reflect upon herself.

The goung girl proceeded in accordance with
the teaching of the healer, and the result was
marvelous. As her mind opened, her lungs also
opened, and she breathed freely. Her changed
mental condition seemed to be reflected not only
in her body, but in her circumstances. From a-
mong those whom she had sought and ministered
to, was a young girl of tastes similar to her own,
in whom she found the congenial companion and
friend for whom she had always longed. Her
asthma became a thing of the past because its
corresponding cause in mind had been discovered
and destroyed.

Now there are many other and more evil conditions of mind that might equally correspond

with the asthma such as avarice or any intentional holding of good to one's self.

A mother came in great distress of mind to a healer, with her child, a boy seven years old, to have him treated for an eruption upon his skin. She said that he had been troubled with the ailment from his birth, and no medical doctor had been able to do more than afford temporary relief, after which the ailment would break out worse than ever. There was no impure blood in her family, and she did not know how her child came by such a taint.

The healer questioning the mother concerning the child's disposition and habits, learned that he had a violent temper in which he indulged on the slightest provocation, and that he seemed to have an affinity for profane and obscene language which he readily caught and remembered.

The mother had never attempted to correct these evil habits in the child, because as she said, he was not to blame for them. They had been put upon him. During several months before his birth, she had been in the habit of sitting by an open window that overlooked a lumber yard, in which were employed some quarrelsome men who

made use of such profane and obscene language,
that, at times she was obliged to close the window,
preferring the air of a stifling room, to the greater
annoyance of those harsh and angry voices.

She was so mistaken that she thought she was
dealing kindly by her child in not correcting in
him a tendency to evil for which she did not consider him responsible.

The healer at once undeceived her. She pointed out to her the correspondence of the impuri-
ties breaking out upon the child's skin to the im-
purities in thought and language breaking forth
from the child's mind. She gave her distinctly to
understand, that so long as the creating evil remained in the child's mind, there was no system
of medication that could purify the child's body.

The healer treated the child for heredity. She
also went still further back and treated the mother
for the evil which she had permitted her mind to
receive and to hand down to her child. She pic-
tured the mother sitting by that open window in
so calm and trustful a condition, with her thoughts
so concentrated on comforting themes—on the
beautiful possibilities of her child's life, and on
the all-protecting power of spirit—that the turbu-

lent voices outside made no impression whatever upon her mind.

The mother's mind being free from disturbance, no disturbing influence could gain access to the child.

It is not that which is presented to the senses, but that which the mind receives and retains, that does the harm.

The effect of the healer's earnest and persistent treatment was shortly manifested in the child. His eruptive temper disappeared and his bad language dropped away from him, consequently the poison which these evils had created in his blood was carried off in natural channels, and his skin became smooth and tinged with a healthful glow.

We know of a lady who desires above all things to have money. She feels that every ill in her life is brought about by a lack of money. She regards money as the oil that so lubricates the machinery of life that with it nothing could possibly go amiss. When she lost her children she felt that if her means had permitted her to call in more skillful medical attendance, their lives would have been spared.

A lack of money is the greatest cross she could

have had come into her life, and it is just the cross that her own condition has brought upon her. There is nothing in the condition of her mind or spirit to attract or build up any better circumstances around her.

You may say that all who have money are not spiritual. Certainly they are not. But those who have money are in a condition to attract money. There is something in their development that brings it to them, but it may mean happiness, or or it may mean misery to them, according to their condition.

The lady to whom we refer, in her state of longing and discontent, has, as might be expected, no physical health. Instead, however, of discerning the cause of her decline, shs feels sure that if she had money enough to seek other climates and scenes, etc., her health would be perfect.

Her ailments correspond to her mental condition. There is a lack of nutrition in her system, she is thin and pale and her blood is poor, and until she discovers her spiritual poverty, she will, in all probability, keep her material poverty.

Physical blindness corresponds to spiritual blindness.

Physical deafness corresponds to spiritual deafness.

If these ailments are hereditary with us, what we have to do to rid ourselves of them, is to conquer heredity. But if they are of our own creation, what we have to do to rid ourselves of them is to cultivate a *willingness* to see and to hear spiritual truth whenever and however it may be presented to us.

CHAPTER V.

CORRESPONDENCES IN ACCIDENTS.

We think that the causes, not only of diseases, but also of so-called accidents, may be found in a condition of mind.

As there are no accidents in spirit, no conditions of mind that come by chance, there could not well be chance in any of the material results of mental conditions.

All the events of our lives—those whose causes we cannot discover, as well as those whose causes are clearly discernible to us—come to us through the working of law. There is no chance, there is only the working of law in what we call our accidents.

If we hold ourselves in a harmonious condition of mind, we are not in a condition to bring acci-

(56)

dents upon ourselves, and we are not likely to meet with accidents.

The mind of man presides over the material things that he presses into his service. We all know that material things serve us well in proportion to their adaptability for our needs and our proper use of them. They are adapted for our needs in proportion as there is mind, that is good thought, put into them ; and we use them properly in proportion as our mind is in so harmonious a condition that we have an unhindered use of the faculties brought to bear upon them.

It is mind that adapts material things for our use, and it is mind that enables us to make the best use of them.

We each have our days and our seasons in which material things do not serve us well, when we know that the material things themselves have undergone no change whatever. The difference is in ourselves, in our condition of mind.

A boy, for example, uses the same sharp knife that he has whittled with hundreds of times and never cut himself. But to-day, being in a disturbed condition of mind—perhaps from a feeling of rebellion or of disappointment—he does not bring

to bear upon his work the calm and consequently
best use of his faculties, and he cuts his finger.
You say he has met with an accident.   But is the
occurrence a chance, a mere happening?   Does it
not come through the working of law just as much
as any other event of life?   Is not the cause of
the boy's accident to be found in his own condition
of mind?   If a horse turns into the wrong street
because the driver has dropped the reins, we do
not feel that the occurrence was a mere chance,
but we logically trace it back to the man's having
resigned his command over the horse.   So with
the boy, when he cuts his finger his necessary faculties are not in command over the knife.

If we meet with accidents it is because we are
in a condition of mind that renders us liable to
them, and they will generally be found to be such
special kinds of accidents as correspond to our
special condition of mind, and moreover just such
kinds of accidents as we require to help us to correct that special condition of mind.   They will be
the natural results of the condition working for our
good in the destruction of the condition.

A lady sprained her ankle by walking so far
on the outer edge of the sidewalk that her foot

slipped off. She said that there was no reason for walking where she did, as no one was passing by, and that she never would have done so, if she had not that day, been in a wavering and foolish condition of mind, and that her feet, like her mind, were governed by no fixed purpose.

One evening, a lady walking briskly along, fell flat upon the pavement, with no apparent reason for so doing, and seriously injured herself. She said that all that day she had been feeling defiant and self-asserting, and it seemed to her that instead of falling in a natural way, she had been dashed down to the ground by some outside power. She was habitually defiant and self-asserting and had many times before in her life met with similar accidents.

Another lady who had for days been in a defiant condition of mind that led her to be determined to pursue a certain course, whether right or wrong, started in a specially defiant moment down a flight of stairs over which she had been in the habit of passing many times daily, and pitched headlong from the top to the bottom. There seemed no more reason for her falling on that day, than on any other day unless that reason was looked for in her condition of mind.

A fall was just the sort of accident that her condition would render her liable to. It was also the most offensive accident that she in her self-asserting mood could have met with, and it was at the same time the most salutary one as it tended to correct in her the error of self-assertion.

We think that one who is self-sufficient, or self-asserting, or self-opinionated is in just the condition of mind that would expose his material body to a downfall. His trust is in himself and not in the only protecting power. While such an accident would be the one most offensive to him, it would be the very kind to which he would be liable, and that he would need to raise him up into a higher condition.

We think that one in a harmonious condition of mind, one with little thought of his own importance, would not be likely to meet with an accident in which he would be thrown to the earth, or if he did so, would not be likely to be injured.

A man in that condition of mind that he felt himself standing loftily above his fellows, and who always managed to stride before others and be in the advance, met with an accident that resulted in a painful and so-called incurable lameness, which he believes will leave him all his life limping in the rear of his fellows. Spiritual science would tell him that his ailment is not incurable, but that when he traces its cause to a corresponding condition of mind, and corrects that condition, all ill effects of his ailment will disappear, even though an accident may have been its secondary cause.

We, like nature, are able to recover from our accidents. We have only to seek and apply the true remedy.

Can we not for further illustration of the working of this law of correspondences, imagine a pugilist, who considers his brawny arms the highest potency in the universe. After a special indulgence in cruelty, in which all his better nature seems paralyzed, he is seized with paralysis of the right arm. Now the dead condition of his affectional nature has caused him no uneasiness whatever, but the dead condition of his muscles makes an appeal to him that awakens his mind to a better condition.

We know of a lady, who, after meeting with disappointment after disappointment, cross after cross, pecuniary loss after loss, until it seemed that she could endure no more, had, as a culmination of her woes, part of her limited, and so-considered necessary wardrobe, stolen from her. Her friends saw only cruelty in the accident. But she has since told us that it was worth more than a fortune to her to lose those garments, because it gave her trust in God.

Now it may be difficult for some of us to see how it would give us trust in God to meet with loss.

But this lady, although she had in a certain way believed that whatever comes to us is the result of our own condition, and that we will be taken care of if we trust in the Infinite Spirit, yet she had allowed herself to be anxious and fearful concerning her affairs, and had not trusted in God, even when she knew that she was doing her best. The accident, which she felt to be the outcome of her own fearful condition, awakened her to a full realization of the importance of trust. She knew that, as every one of our misfortunes has a lesson in it, she had only to seek and learn the lesson conveyed to her in her loss, and she did so. She learned that unless the Lord keeps the city the watchmen watch in vain. She learned that it is not fear and anxiety, but trust in the loving Father's care that protects us from even worldly adversaries.

We have known of a number of persons whose trust has been in themselves and in material things, who have several times in their lives met with a serious loss by robbery, etc., of these very material things upon which they set an undue value. Now, like Shylock in Shakespeare's drama, they have been in just the condition to bring upon themselves such a loss as was more grievous to them than any other accident could have been, at the same time it was just the kind of accident that they needed to raise them to a higher level.

I think we all feel that in order to appeal to Shylock to work a reformation of character it would be necessary to deprive him of his money which was his God. We all know that those who fear dishonesty and look for it, generally find it. Those who watch for thieves will attract them, and be rewarded for their diligence by finding them.

Fear is a complete inversion of trust. It is a belief in the power of evil, in the supremacy of evil over good.

Fear places us in that incompetent condition of

mind in which we are playing into the hands of evil.

A trapeze performer and a tight rope dancer know very well that without fear their footing is sure, but that the moment they fear they are lost, and all this with no change whatever in the material things with which they are dealing. A confident condition of mind, in which they can command their necessary faculties, is their only salvation.   ,

Fear of a certain evil attracts that evil to us. Not only the diseases we fear are likely to come to us, but the accidents we fear are just the ones to which we are most liable.

We have known persons who were in constant fear of death by accident, who have at last met with such a death.   Fear creates, encourages and attracts evil.

It is easier to trace to our condition of mind our own personal accidents, than those public calamities for whose occurrence we seem in no way responsible.

We may, for example, be able to see that the breaking down of a bridge under a railroad train, is the result of a wrong condition of mind in those who have charge of the bridge and should see to it that it is kept in good repair; but we are perhaps not able to see how we, who are only a passenger in the train, could trace *our* share in the accident to *our* condition of mind.

Nevertheless we believe that one in an habitually trustful condition of mind, would, in some way, be protected from harm by accident even of another's creation.

A man who, in crossing the railroad track vascilates before the coming train, turning first in one direction and then in another until it is too late and the engine is upon him, would not, if he were in a harmonious condition of mind, in the first place, be likely to cross a track until he knew that no train was near; but, in the second place, even if he were in that respect in some way deceived, he would not, when he saw the train, be likely to be so overcome with fright that he could not command his faculties. His well regulated mind would be likely to serve him well in the emergency, and with lightning speed he would fly in one direction and be saved.

One whose soul rested in trust would, like the three holy Jews of old when cast into the fiery furnace, be delivered by the God in whom he trusted. He would, like Daniel, when cast into the den of lions, although exposed to the greatest danger, yet, on account of his condition be saved from harm.

It is said that God sent his angel and shut the lions' mouths. Has not God his angels to-day just as much as ever he had, even if we do not see them ? Has he not his ways and means of protecting us even though we cannot look into the spiritual realm and see just what they are?

How many cases of miraculous escape from accident we all can recall. How many cases we hear of, in which persons have been in some incomprehensible way hindered from exposure to an accident which would seemingly have destroyed their bodies, or they have been brought through the accident with little or no injury, while others around them, in apparently less peril, have perished.

Are these occurrences the result of chance? There is no chance. We must refer them to the working of some law. The Old Testament, as well as the New Testament, is full of illustrations of the working of this law, in which those who trusted in God were saved from harm.

This saving trust does not come to one who is not living in harmony with divine law. Although it is a condition of mind, yet it cannot be commanded at a moment's notice. It must have become habitual. It is born of overcoming self and of the doing of good works.

The whole life of Jesus was an illustration of the working of this law. For him, who overcame the world, that is, who rose in perfect dominion over his lower nature, there was always a way of escape from harm. He was able to take himself away from the pursuing rabble that threatened him. If his boat had pushed out into the sea and left him behind, he was able to walk upon the waves and reach it. With all his exposures to danger, it was not until the last act of the sacred drama, after he had decided that it was expedient for him to go hence, that, on giving up his material body to the enemy, he suffered any physical injury. He could then, as he said, have prayed the Father and he would have given him more than twelve legions of angels, that is he could have desired it and trusted in the Father and he would have been rescued. But he saw that the time had come in which he could do a greater work by going than by remaining longer, so he permitted the enemy to deprive him of his physical body.

Have we any of us ever given ourselves the opportunity to learn what it would do for us to go through this mortal life with a feeling of perfect trust in God's willingness and ability to protect us, not only from all manner of disease, but from all manner of accident?.

CHAPTER VI.

INDEPENDENCE.

There is much fine talk in the world about independence. In order to be spiritually independent one must be able to maintain a continued being with no aid from other spiritual beings, and no *necessary* relation to them. In order to be materially independent, one must be able to maintain an existence in the material world with no aid from other human beings of the same world, and with no *necessary* relation to them. But there is no such thing as either of these kinds of independence. It is only in a very limited sense that the word independence can apply to any being, creature, or thing in all God's great universe.

Every theist acknowledges that he is wholly dependent upon his Creator; but every theist does not perceive that much of his dependence upon his Creator is through his fellow beings, as in mechanics, the lever depends upon the fulcrum and the fulcrum in turn depends upon the foundation on which it stands, without which firm foundation the whole machine would be useless.

No one human being can stand alone in spirit or in mind. We are never alone in one single feeling or thought, and as it is feeling and thought that govern our material life, neither can we stand alone and unaided in the adjustment of our material affairs.

The whole universe is one grand concatenation of God's manifestations, each related to all the rest, each a dependency of all the rest. Man is no more independent of his fellow beings than one planet is independent of all the other planets of its solar system, or than one solar system is independent of all the other solar systems of the great harmoniously moving universe. Each individual is but one link of an endless chain, from which he can no more detach himself, than the earth can sever its relation to its sister planets and depart from its preordained orbit.

Since by an unalterable law of our being we

are never one instant alone, it behooves us to see to it that our companions are good companions, that our co-workers are those who do good work. How are we enabled to make the right selection? We have the right of choice in the matter. We inevitably become united to that sphere of thought that corresponds to our own thought. The good or evil in ourselves brings us good or evil companions, and without consciousness or effort on our part.

We should, instead of feeling that we are alone in our pursuit of good, rather feel that by the very pursuit of good we are enlisting ourselves in the numberless army of the angelic hosts, by whose aid we are protected in that good.

We need never fear that in this natural dependence upon other harmonious minds, evil may attach itself to us, for no enemy can possibly effect an entrance into God's army. There is no law by which evil is permitted to conjoin itself with good. There is no guise in which evil can clothe itself, by which it can elude the governing spiritual law of like being attracted to like.

There is a harmful, a monstrous, even a blasphemous pride that leads one to feel that he can maintain his being without a God. There is a more common but less harmful pride that leads one to plume himself upon needing no service from his fellow beings. While we should be willing for the sake of right to stand alone, so far as we on our low plane can perceive, yet we should always know that we are never for one moment independent of God who created us, or of our fellow beings to whom we are indissolubly related.

The feeling of self-sufficiency misnamed independence closes the door of the spirit to that saving faith by which we live. It deprives us of numberless spiritual graces to the weakening of our souls and bodies.

As our feelings and our states of mind govern our conduct, this spirit of self-sufficiency presiding over our external affairs brings inharmony into our associations with our fellow beings, and deprives us and them of the joys of that sweet interdependence intended as a blessing to us.

Of course in a certain limited sense, it is right for us to be independent. We should be self-supporting. We should not lean upon others for that which it is our duty, and should be our pleasure, to gain by our own efforts.

Every child, and so far as possible every animal, should be taught to earn a living in some suitable, pleasurable, and healthful occupation.

We do not all of us need to go out into the busy world in order to earn our right to live. We can earn it best by the cheerful performance of whatever duty comes to our hand to do. We have known daughters of rich parents, as well as wives of rich husbands, on whom were bestowed every luxury, still earn it all, every day of their lives, earn their place and right in the lavish household, by their loving and necessary ministrations. The just and harmonious interdependence of such families constitutes their greatest happiness, and by its sacred observance each member fully earns a daily living. In such cases no one member of the family wishes to be independent of the others, except so far as never to shirk a personal duty or responsibility.

We have also known men and women, having no family ties or duties, who have, by their ministrations to the world around them, richly earned a right to live.

We have known poor relations living with richer ones upon whose bounty they were said to be de-

pendent, earn much more than a living every day
of the year.   We have known wives who were
said to be dependent upon their husbands for support, who instead of receiving as their rightful
earnings, the joint salaries of housekeeper, gov-
erness and nurse, have received, from the one who
pretended to love them, only food and shelter and
a begrudged pittance for which they were obliged
to beg.   In such cases are the wives dependent
upon the husbands, or are the husbands dependent upon the wives, and which one of them most
truly earns their right to a support?

There is no word, the true meaning of which
is so little understood as the word independence.

There is no rightful living or rightful place, in
this world or any other world, for one who en-
deavors to sever his relation to his fellow beings
and live wholly for himself.   It is in doing for
others, whether we have a fortune or not, that we
earn our living.

When one has means sufficient to supply his
material needs without the necessity of working
for them, he is said to have an independence.
But is the man who draws his money from his
bank, necessiarily more independent of his fellow

men than he who draws his money in a weekly salary from his employer? You may say that the former owns his time and therefore has leisure for something besides monotonous drudgety. But it is not the *amount* of time we have that benefits us ; it is the way in which we employ our time, be it little or much, that is all important to us.

How frequently we see one, who is in possesion of this independence as it is called, who is so dependent, even in the generally accepted sense, that the day laborer in the streets has more freedom of action than he. His surroundings are so inharmonious, or he is so cramped by unavoidable claims upon him, that his time and even his thoughts seem never at his own disposal. Or, again, he may have his whole time, and not know anything what to do with it, not put it to any profitable use.

We can easily find those who employ their one hour a day to greater advantage, than others their whole twenty-four hours.

Then, in the very highest sense, we all of us have our twenty-four hours a day. Our employers do not control our souls. No employer can hinder us from so possessing and ruling our own

spirits and minds that we accompany our daily
work with the very highest feelings and thoughts
of which we are capable. Who can say that,
under any circumstances, he has no time of his
own, no time for self-culture.?

Instead of calling money an independence, we
should call harmony of spirit an independence.
Harmony with divine law is the only freedom there
is, and it is a freedom which enables us to perceive and to enjoy that sweet dependence upon
our Creator and our fellow beings, which is our
rightful and rich inheritance.

To receive from others with a perfectly right
feeling is a grace of spirit rarely met with. To be
pleased to give others the pleasure of doing for
us, without permitting our pride, on the one
hand, to be on the alert, lest our self-sufficiency
be impugned ; or our greed, on the other hand, to
become excited so that we selfishly desire more
than is right for us to accept, is to attain a divine
medium, to stand on that central point, upon which
alone one can become poised in the rectitude of
true dependence.

We know a lady who is sensitive about having
the price of anything reduced for her, to suit her

reduced circumstances. Of course this feeling is less common and less reprehensible than its opposite, which we see in others who are so in the habit of getting things at a reduced price that they feel defrauded if they are ever obliged to pay a just price for anything. Nevertheless this feeling of independence, as it is called, is founded on a false pride, a false delicacy. It is well to be so imbued with justice that we desire to give a just equivalent for everything we receive whether any one knows of it or not. Justice should be our ruling motive. We will dissect an instance of sensitiveness in the lady to whom we refer, and see of what stuff it is made. When her physician, knowing her limited means, reduced his fees for his services to her, she was, as she expressed it, mortified and distressed. Not that she thought that he needed the money, or that she considered his half fees too little for what he had done for her, but she did not want to feel and she did not want him to feel that she could not pay what others paid. She liked to be independent, she did not want to accept favors from him. She was proud of her spirit of independence. If the amount of his half fees had been his full price, she would have thought it quite sufficient and would not have desired to give him one additional farthing. She continued to indulge in the tormenting reflection that she had not been able to pay his price as others did. This false pride was an obstacle in the way of her recovery to health. Instead of seeing her error and correcting it, she morbidly permitted her mind to cling to it, and her convalescence turned into a settled invalidism.

If we would carefully examine our unhappy conditions of mind that we consider the result of right and even heroic feelings, we would be astonished to find that they were made up of vice instead of virtue, or at least error instead of truth.

We heard a gentleman say that he was perfectly independent, that when he received a favor he repaid it at once so as not to be under obligations to any one. He talked inflatedly about not being beholden to any living being, when the very roof that covered his head, and the very food and clothing that ministered to his body, to say nothing of the thousand and one superfluities that met his fancied requirements, were all supplied him by the skill and industry of others.

In what way was he self-sufficient or independent? Paying for such time and material as he received from others, was merely honesty, and it seemed that even this honesty was not founded on the right principle, for instead of wishing to give an equivalent for what he received because it was only just and right to do so, his prominent thought was to be self-sufficient, to sever that natural tie of dependence intended to foster in us gratitude and love towards our neighbor.

He was proud of his spirit of independence. He thought that all obligations could be canceled by material things. He felt that his money could pay for everything he received from his fellow beings.

Money is an equivalent for only material things.

Whatever good thought or good feeling was bestowed on that gentleman could not be paid for in money. When he paid his hatter for his hats, and his tailor for his clothes, and his shoemaker for his shoes, he paid for only time and material. Whatever conscientiousness, good inrention, good thought, was put into the work of these artisans, could not be paid for in money.

Good thought and feeling belong to the spiritual kingdom and have no equivalent in material things.   Money does not enter into the spiritual realm at all.

There is only one way in which to repay spiritual favors, and that is in gratitnde and loving consideration.   The gentleman to whom we refer would have been offended at the bare thought of loving his shoemaker, for example ; yet would not a shoemaker come under the head of the neighbor whom we are commanded to love?   We are commanded to love our neighbor because it is best for *us* as well as for that neighbor.

This gentleman's condition of mind was incompatible with gratitude or loving consideration for others, which is the only possible payment for spiritual favors, therefore with all his money and all his boasted independence, he most signally failed to repay the favors that were constantly bestowed upon him from the world around him.   Any one of his so-considered dependents may, in this high sense of the word, have been more truly independent than he.

He was not, and as we have eudeavored to prove, could not be self-sufficient, and his very endeavor to be so, his struggle against a natural, a divine law, was an injury to him that reflected itself in his physical body in a corresponding malady. His body became enlarged with dropsy. His self-hood was increased by an appropriation to himself of error; his members were increased by an appropriation of a foreign secretion. The inflated condition of his mind was photographed in his body.

One cannot battle with divine law and come forth from the conflict unwounded.

We have known those who were so grounded in this false pride, misnamed independence, that they could not enjoy the slightest favor until they felt that they had given a full equivalent for it, not in good feeling, but in some material thing. Such erroneous conditions of mind are not conducive to sound health. They are created by an over estimate of self, and the sooner they are destroyed, the sooner the mind will be able to command health for the body under its dominion.

The sooner we are able to perceive, acknowledge. and delight in the divinely appointed relation of interdependence intended to bind all souls together in one mutually loving community, which is still further dependent on the Infinite Being, the sooner we shall be whole and sound in both soul and body.

When we reach a condition of perfect health inwardly and outwardly, then there will fall upon our ears harmonies sweeter than the music of the spheres in which the heavenly bodies are moving on in perfect accord with one another and with their great central sun that rules their sidereal heavens.

CHAPTER VII.

___

___

In what does real success consist? Will our real success in life consist in our reaching a certain point for which we are striving in the belief that at that point lies our happiness, or will it consist in our so harmonizing ourselves with God's plans for us that, by thus employing the only sure means, we have *already* arrived at the only true happiness?

Whatever duty lies at hand for us to do, is a part of God's plan for us.

It has been truly said that all failure is success. Failure is the negative side of success. In our every failure there is a lesson for us, which, if we will search for it and find it and learn it, will so in-

(83)

crease our wisdom as to advance instead of retard us on the road to success.

Failures are only the little valleys between the foot-hills, through which we must pass before we can begin to ascend the mountain; and each little descent as well as each little ascent takes us nearer the grand height.

We may make great mistakes, we may bring upon ourselves the most painful consequences, by our evil or ignorant condition of mind, yet, if whatever consequences ensue to us from our lack of development, work for our improvement, how can they be considered failures?

The only real failure there is for us, is *our* failure to profit by our non-success.

When we fail in any undertaking, we are prone to feel that we have met with a total loss, that all our efforts are without result other than disappointment to ourselves and all concerned.

No sincere effort directed by good intention is ever wasted. If it does not effect the good that we, in our shortsightedness have in view, it is sure to effect some other good. Even if our efforts are directed to the attainment of some selfish end that we fail to compass, our failure teaches us the very lesson we need to learn and thereby helps us on in the working out of our salvation.

We have heard men who have met with failure in advanced age, say that the work of their life time was destroyed. The good work of a life time cannot be destroyed. The results of our life work are so fixed upon us that they have become a part of ourselves and we must carry them with us wherever we go.

If we would not have failure let us not need the lesson they are intended to teach. Let us not be in that spiritual and mental condition, which, by the working of divine law, brings these failures upon us. Let us, in going over life's foot-hills, so bridge over the valleys with wisdom that we can pass from one little height to another without these descents which we call our failures. If, however, we are not just yet able to do so, let us not be discouraged, but always look to the next hill beyond, feeling sure that we are not only going onward, but upward, so that finally the very valleys we enter are higher than the hills we have left behind us.

If we fail in any undertaking, there is a reason why we fail, and that reason is within ourselves.

There may be secondary causes in the circumstances around us that seem to bring us failure, but the primary cause, which built these circumstances, is in ourselves, in our peculiar spiritual and mental condition.

Napoleon Bonaparte was right when he said that we make circumstances; but he was mistaken in his conception of the *way* in which we make them. It is not done by our personal will, but by our spiritual condition, as he had an opportunity of learning at the close of his career.

Until we come into the highest condition, until we are fully illuminated, we must expect, like one mounting a dimly lighted stairway, to stumble now and then, we must expect what we call failures, both in our spiritual strivings and in our material efforts.

here are so many who eagerly grasp at any suggestion that may aid them to material prosperity, but they do not care to learn anything concerning the road to spiritual prosperity. They do not know that there is but one road to both, that the latter includes the former, and that if they seek the greater they will be sure to possess also the lesser.

It has been argued that it would be fatal to our material prosperity to devote ourselves exclusively to our spiritual condition, leaving our worldly affairs to take care of themselves, sitting down and passively awaiting the turn of events.

We would reply that such a view of our duty would be a total perversion of the teachings of spiritual science. There is nothing fatalistic in our line of life, as foreknown and preordained by the great Omniscient Being. We are not to drift with the current rudderless and oarless. We have our share of work to do in connection with each event of our lives. Within certain limits we shape the course of events for ourselves. A true spiritual scientist will, in even the least important of his affairs, employ every means in his power to bring about a legitimate result, and, working, as he will work, without inordinate desire, or anxiety concerning results, he will be in a condition to discover and to select the most effective means to accomplish a proposed end. Instead of being indifferent, and letting his worldly affairs take care of themselves, he will take better care of them than a materialist could possibly do. He will trust in spirit, the only true power, and he

will thereby have a better use of all his faculties.
He will be wiser and more far-seeing.

We know a man, who is governed by an inordinate desire for worldly success. His business is a
rational one, and his working for success is perfectly legitimate, but with all his efforts he sends
forth so much anxiety and so many fears that the
success he seems to have reason to expect does
not come. Every fear and every anxious thought
that he sends forth into the realm of mind works
against him. For example, he fears that a certain
man is going to be dishonest with him, he thereby
sends to that man thoughts of dishonesty, which,
if he has any affinity with such thought, enter into
his mind and aid in the generating and furthering
of a dishonest plan. His fear and anxiety not
only cause him to play into the hands of the
enemy, but they bring his own mind into that disturbed condition in which he has not the best use
of any of his powers. His memory becomes weak-
ened, his judgment becomes clouded, his foresight
and forethought depart from him, and he becomes
daily more incapable of discerning the most effective
means to ensure the success for which he is ready
to sacrifice every other good.

Of course all this tells upon his physical health, which is broken down. He looks pale and haggard and is nervous and irritable in disposition. If any one should tell him that his own condition of mind was the only obstacle in the way of his success, he would not be able to understand it, yet such is the case. There is nothing resembling trust in his soul. Even when he is doing his very best he has no faith in the good result of his efforts.

Inordinate desire generates fear. Fear is a complete inversion of trust, and is destructive of mental tranquility and of physical health.

Too much and too incessant effort frequently injure an undertaking. It indicates greater ability to be able at times to do nothing than to employ gigantic effort. The majority of business men feel that the issue of every undertaking depends wholly upon their ceaseless activity and their personal will power directed to material things. Activity and will power when wisely directed to spiritual endeavor are powers worthy of consideration; but to direct any of our powers wholly to material things, leaving out the causal realm of mind, is eventually to fail in even those material things.

We should take up any occupation with an intention to do our best and to succeed, leaving out all anxiety, and fear for results, and sending forth kindly and encouraging thoughts to employer or employed, then, if success does not crown our efforts, we may receive it as an indication that some better occupation awaits us, which by patient endeavor or perhaps by only patient waiting, will be sure to come to us.

It is only by pursuing such a method as this that we can learn what our life work is to be. Do we give ourselves an opportunity to learn what success this method would bring us? How many of us are wholly without anxiety or fear concerning any result whatever? If we are not in this high and trustful condition, then our first and most important work is to endeavor to place ourselves in it. While we should always intend to do the best work of which we are capable, we should think more about maintaining a harmonious condition of mind than about our work. Interior harmony, with a right intention, will do more than the most intense desire without this harmony, in composing a fine piece of music, or painting a fine picture, or writing a good book, or building up a large business, or in any work whatever. If we work harmoniously our work is sure to be good work with little thought about its quality.

Why is it that out of the vast number of men and women, who start a business on their own responsibility, so few succeed in their undertaking?

In the first place, very few of us give ourselves the opportunity of learning what our work really is. Influenced by vanity, pride, or ambition, or it may be by the less reprehensible motive of better serving those dependent upon us, we take up a business for which we are not adapted. The result is that our vanity or pride or ambition receives a shock, or those dependent on us lose instead of gaining, by our utter failure in a worldly point of view.

Failures are frequently the result of a lack of ability in the required direction. But every human being has an ability for something, and he can find that something if he will keep his mind in the condition that permits it to reach him, and permits him to recognize it when it comes.

We do not know all the unseen forces that are incessantly at work for us. If we leave ourselves more passive in the hands of good, we shall find ourselves more wisely directed than we could possibly be under the guidance of our own selfish inclinations. It is just as impossible to tell what ability a man has, when the wrong motive guides him in the choice of an occupation, as it is to tell how well or how fast a car can go when we see it only off the track. Whenever any of our selfish desires guide our conduct, we are off the track, and we need not be surprised if we, not only receive an unmerciful jolting, but also never reach our destination.

In endeavoring to find our own peculiar work, we should begin by doing faithfully the work that lies at hand. It may be that we need the discipline of doing much work that we neither like nor excel in, before we can be prepared to take up the work we enjoy doing and in which we shall eventually succeed. If so, let us equip ourselves with patience and courage, and thereby shorten the term of our probation and render ourselves stronger for the congenial occupation when it comes.

How many square men try to fit themselves into round places? It is not because square men like round places, for quite the contrary is the case. A square man is naturally attracted to a square place. In our natural, our best condition, we are attracted to the sort of occupation suited to us. If we are actuated by a desire to do only what is best, we shall be sure to find that best. Regard for a high position, and other worldly considerations impel us towards a certain vocation, and we have but the one thought, the one desire, to reach it at all hazards. We reach it, we pursue it, and we fail in it.

For illustration, two young men started in business at the same time, on opposite sides of the street. One opened a shoe store and the other, a painter's studio. They have been in business some years and are neither of them successful.

The painter displays the very best judgment in attending to the pecuniary details of his business. It is a pleasure to go forth and buy his materials. He enjoys every part of his business except the painting itself. He is fond of opening his studio to the public and arranging for sales, but unfortunately his pictures are poor and the expected sales do not take place. He is annoyed and discontented and wishes he dealt in a stock that would sell rapidly. He envies the young man across the street, and feels sure that if he had that shoe store he could build up a business in no time, for he knows all about tanneries and leather. He wanted to be a merchant, but he, with all his family, thought it would be such a fine thing for him to be a painter, that he determined to be one.

The proprietor of the shoe store spends every spare moment in sketching. A lady customer comes in and he is so struck with her picturesque face that he pays little attention to her wants, and she determines never to go into that store again. He is cheated in the leather he buys. He knows so little about his stock that he does not discover the dishonesty of a clerk. There is nothing in his condition of mind to attract or to keep customers. He tries to be conscientious, but he feels such an indifference if not aversion to his business, that his mental atmosphere keeps it stagnant. He has a talent for painting and he always wanted to be a painter, but the uncle, who set him up in the shoe business, would not open a studio for him, and, feeling that he must not throw away pecuniary advantages, he made the great mistake of fixing upon the wrong vocation.

Now, if these two young men could have changed places, undoubtedly they would both have been happy and successful.

If the painter had thrown aside all selfish considerations he would have been able to see clearly what his vocation was, and he would have had the courage to pursue it.

If the shoe merchant had refused the start in a business for which he was unfitted, and taken up some temporary employment, and trusted and waited patiently, his time would have come to pursue the vocation for which he was fitted.

We know of a talented painter, who in his early days had not the ghost of a prospect of ever being able to follow this calling. But he patiently and trustfully waited his time, not in idleness, but in industry, and he is now, while still in his youth, established in a fine studio and doing well.

We can learn what wonderful things God can and will do for us, only by pursuing the right and putting our trust in him.

Instead of feeling that ambition and ceaseless activity are required to push us into success, let us realize that the conscientious fulfillment of the duty of the hour, whether it be working or simply waiting, can alone start us on the road to true success.

When women as well as men bring themselves into so harmonious a condition of mind that they desire to follow the vocation for which they are best fitted, that vocation, whatever it may be, will be open to them. They cannot be kept out of places for which they are eminently fitted.

We say to all women, instead of complaining that you are not allowed to enter certain fields of usefulness, fit yourself for those fields—and you cannot fit yourselves for them if your work does not lie in them—and you will then by a spiritual law be so attracted to them that no one can possibly keep you away from them.

If we are in a harmonious condition we will not be likely to undertake anything that is not suited to us, we will not be likely to attempt anything that is irrational, for spirituality, instead of conflicting with rationality, always includes it. Whatever we, in a good spiritual condition, undertake, even though it may bear upon this material existence, will be quite likely to succeed. All our plans will be modified with an *if*. We wish to succeed *if* our lands are God's plans, but *if* they are not, we are perfectly willing to have it proved by the total frustration of them. No inordinate desire can take possession of a harmonious mind. When we rise up into a high condition we shall not need to think so much about this *if*, for we will naturally and unconsciously co-operate with divine intention. God's plans will be our plans.

The members of an orchestra do not each render their strains in accordance with their own ideas of time and expression ; they follow their leader. Did they not follow their leader, discord would take the place of harmony. When a musician is unpracticed he must closely watch the *baton* of the leader ; but when he becomes educated and skillful, he will be so in sympathy with his leader that he will scarcely know he is being led.

When we rise up into harmony, we are in so close sympathy with our Divine Leader that we do not know we are being led. We are one with him. As Jesus said of himself, " I and my Father are one," so we each can say when we in purpose become one with the Ruling Spirit of the universe.

Then, and not until then, shall we have true success.

CHAPTER VIII.

OLD AGE.

What is old age? When we have passed a certain number of years in this life, what part of us is old? We surely cannot suppose that even scores of years count as anything for the immortal part of us that lives forever. It must, then, be our material bodies that we consider old. But are they in reality ever old?

When we regard material objects with our material senses alone, our ideas concerning them are not only limited, but frequently false. The earth which moves so fast seems to be standing still in space, while the sun, which stands so firm in its system, seems to be ever moving around us.

A mass of iron and a mass of rock, which seem to us so solid and so still, are by a finer sense dis-

(98)

covered to be composed of non-contiguous moving atoms.

There is in reality no such thing as still life. All is life and life is motion. Spiritual life pervades the universe. All matter is projected and sustained by spiritual power.

What we call inanimate nature is not only full of motion, but it is full of sound. If the finer vibrations of moving matter are imperceptible to our gross senses, the limitation is in us. If our dull ears fail to catch the vast number of tones above and below those few octaves of which we have knowledge, the limitation is not in nature's instruments, but in us. If to one the flowing water of the brook sends forth a full clear song, while to another it is only motion without sound, the difference is in the varying perceptions of the two.

To gain an accurate knowledge of even matter we must employ a perception higher then the physical senses. Matter is ruled by spirit. To gain a true knowledge of matter we must learn something of spirit.

As inorganic matter is held in form by spiritual power, so organic matter is changed in form by the same power.

Inorganic matter moves, but does not grow. It is not recuperative. Organic matter moves and grows and is recuperative.

Inorganic matter is nature's store-house of raw material, while organic matter is the workshop that, with a definite intention, selects and utilizes the raw material.

Nature's store-house as we, with our physical senses perceive it, is still and lifeless, while her workshop is moving, busy, and full of life and intention.

The most wonderful piece of organic matter, our material body, unlike a piece of wood or stone, is constantly changing its atoms. It is incessantly throwing off those for which it has no further use, and from nature's lavish store, accreting to itself new ones.

We say the body does this for itself, but the body no more performs these operations than the raw cotton in a factory weaves itself into a fabric. It is mind that presides over all working in matter.

The work that is carried on in any shop, is presided over by the mind in charge of that special shop.

Our material body is our material workshop·

We are placed in special charge over it. If the work therein executed is not done normally, that s, in accordance with divine law and intention, it reflects on us who have charge of it, and we are just as responsible for its condition when we have been in charge over it for ninety years as we were in the beginning.

Let us endeavor to see this body that we consider so solid and so real, as it truly is, as made up of myriads of moving atoms that do not even touch one another, and that are held together and at the same time separated by a spiritual intelligence, which is constantly sending them forth and taking from the elements new ones to fill their places.

That illustrious member of so many medical fraternities, *Mons. J. M. Charcot*, as well as other scientists, advances the opinion that we have an entirely new body every year. *Mons. Camille Flammarion*, another celebrated Frenchman, says this entire change takes place in less than a year. Others, who lay claim to a finer and more accurate perception concerning the subject, assert that we have an entirely new body every few months. There are those who have been bold enough to advance the idea that one month is sufficient to work this entire change.

How solid then is that body which consists more of space than of molecules? How real and lasting is that body, which is never the same during any two consecutive moments? How old is that body, which in the light of ordinary science, has never at any time, in any of its atoms, been in existence more than one year.

We would not talk so much about the body being old and wearing out, if we fully realized that it is never more than one year old.

Why should we suppose that spirit, after it has sustained and renewed this body for a certain length of time, becomes tired and can no longer do its work well? Spirit is never tired. Spiritual power is never exhausted or even diminished. So long as spirit has charge of a body it is able to keep that body in a sound condition, if we permit it to do so. Spiritual power, when permitted to flow in upon us, will continue good work just as long as that work is needed.

It is for us to see to it that we give this vital force fair play, to see to it that we do not hinder the divinely appointed work.

The body under our dominion is like clay in the hands of the potter. It waits for us to bring to bear upon it the fire of truth and give it a more perfect quality.

It is neither to be expected nor desired that we should keep this material body forever; but so long as we do have it we should keep it in good repair.

It is not to be supposed that even with the most perfect harmony of mind, the body will undergo no changes in appearance. Any change in the mind will produce a change in the body under its dominion. As we do not expect to know of a mind that has undergone no change from infancy to old age, so we do not expect to know of a body that has remained the same throughout an earthly experience. But change does not necessarily mean disease.

A plant is not subject to blight by reason of its age, but by reason of something wrong in the conditions that govern it. So it is with man. Disease comes upon him, not by reason of his age, but by reason of an inharmony in the condition of mind that controls his body. The grand old Sequoia trees, that are estimated to have stood for four thousand years in the forests of California, and that now rear their verdant branches up over three hundred feet into the blue ether, evince no more signs of blight than the young saplings that have just sprung up around them   The texture of their wood and of their bark is not the same as that of the wood and bark of a young tree, but these changes do not include blight or decay.

One at eighty years of age would not, even under the most perfect conditions, present the same appearance that he did at twenty; but he might be even more sound and healthy. The youth of twenty does not look as he did when only one year old, yet he may be even more sound and healthy and free from suffering.

Why is it, then, that we decline with our advancing years? It is because, when we do not create disease, we harbor it. We believe, and all the world around us believes with us, that a certain class of ailments is an inevitable accompaniment to old age, and this belief invites the ailments to come upon us, and nourishes and perpetuates them

Instead of yielding to this universal error, let us feel sure that there is no ailment whatever that need come upon us because we are old. Let us

realize that under the dominion of a mind in harmony with divine law, the body, in the incessant
work of recuperation that is going on, will accrete
to itself only sound particles, and moreover that it
will adjust these particles in accordance with the
same law and order that governed the formation
of that body in the beginning.

Whenever health follows disease, it is because
unsound atoms have been thrown off and sound
ones have been taken in their place, it is because
misplaced atoms have been normally readjusted, it
is because functions have been set normally to
work again, it is because of any one or all three of
these changed conditions of the material organism.
Let us then, by an exercise of our faith, by an
effort of our will and imagination, invite what is
called nature, not only to set about this good
work, but to continue it indefinitely.

We assert that in old age we can, not only retain
good health when we have it, but even if disease
is upon us, we can drive it forth by the self-same
method by which we drove it forth in youth. If
our body of this year, irrespective of our age, is
diseased, we can, during the work of re-formation,

so control the new body of next year, that it will be a sound, whole body.

Disease does not perpetuate itself. It is kept alive by conscious or unconscious mind in dominion over it.

Why is it that scars upon the skin and flesh are reproduced indefinitely? Not a single molecule of that skin or flesh that forms the scar is the same as at this time last year, and yet there is the scar the same as ever. It is a faithful copy of last year's scar. It has appeared exactly the same again and again, for perhaps the last fifty years. Our mind, together with other minds, has retained that model and externalized or copied it in our body fifty times, and just so long as it *is* retained in mind, it will continue to be externalized in matter.

Nature, that is, spiritual law working in matter, will continue to copy an old model until a new one is presented. When we destroy an imperfect model and set up a perfect one in its place, then we may reasonably expect that our bodies will present a faithful copy of that perfect model.

Now, although it may not be worth any great concentrated effort to rid ourselves of scars, since they do not seriously interfere with health or happiness, and we have higher uses for our higher powers, yet it is certainly worth great effort to rid ourselves of unsound organs or members. If, for example, our lungs this year are ulcerated, we should not hold these ulcers in mind so that they cannot avoid being reproduced in our new lungs of next year. On the contrary, we should be willing to make any amount of effort to correct the cause in mind—which may be grief or anxiety, or any other erroneous condition—and also to form a new model of healthy lungs, which shall cause all the unsound atoms to be replaced by sound ones, forming in our lungs a photograph of our perfect thought; and this can be done at any age.

It is said to be perfectly natural for certain mental and physical faculties to fail us in advanced age. Spiritual science would assert that there is nothing natural about it, that there is no reason why any faculty, mental or physical, should, with continued exercise, ever, at any age, fail us. All our faculties are made for use, and if we defeat divine intention by not using them, they are quite likely to fail us. If, for example, we place one of our arms in a sling and never use it, we pursue a course that will render it useless. So it is with

any of our physical and mental faculties.   We
need not expect to have in old age the normal use
of any faculty that we have allowed to rust.

It is a widespread error to expect our eyes to
fail us in old age although we may have made the
wisest use of them all our lives.   Why should they
fail us?   At any age whatever our eyes are not
more than one year old.   If we are ninety years
of age we have had ninety pairs of eyes.   We do
not expect an infant one year old to have failing
sight.

You will perhaps argue that the lens of your
eyes becomes flattened and hardened with age; but
as we have endeavored to show, your eyes have
no age, they are always new and young.   Of
course if the lens of your eyes is flattened and
hardened, you will have defective sight; but why
should it be so, or if it is so, why should it remain
so?   It must be because our mind, as it advances,
and should become more capable, permits itself to
do poorer work, so that each new pair of eyes
constructed under its dominion is a little poorer
than the last pair, until we have, what we, in com-
pany with all the rest of the civilized world expect
to have, viz.: failing sight.   We do not intention-

ally do poor work. We do not intentionally shut out the spiritual life force that is necessary for a normal reconstruction of our material bodies; but we are so governed by prevailing false belief respecting age in our material body, that we cannot retain a perfect model for the work that is going on. In this way our eyes, as well as every other member, are subject to a constant maltreatment from us and from all the world around us. How can we expect them, having no knowledge or power of their own, to resist these maltreatments?

If a drawing master should set up upon his platform a deformed figure for his pupils to copy, he would not be surprised to find the deformities of the figure reproduced in the work of each pupil Models of the human figure as deformed and diseased, are already set up for us in the realm of mind. They have been set up and fixed in the world of thought for ages. It is not, therefore strange, that we continue to reproduce them. Let us tear them down! It will, of course, require effort, but effort against error is rightly-directed effort. Let us vigorously deny all these universal errors regarding our failing faculties. Let us confidently assert that, so long as spirit has charge of reconstructing our eyes, for example, just so long spirit is perfectly able to do its work well.   Let us rest assured that, whatever be our age, there is no reason why the new atoms, which our eyes are momentarily accreting to themselves, should not be perfectly sound atoms, and further, that there is no reason why these sound atoms should not be ad usted in strict accordance with the law that governed the formation of all our new eyes during the years of our youth.

If the work of forming new eyes is normally performed, as it will be under the dominion of a mind in truth, the lens cannot become flattened and hardened.   If it has already become so, then let us apply the only sure remedy, a mind in truth. and it cannot remain so.

Why is it that there are numberless cases on record in which persons of advanced age have experienced no failure of sight?   Whenever such a thing occurs, it occurs in accordance with some law, for there is no chance.   If a thing occurs once it is proved possible.   Whatever is possible, can, under the same conditions, occur again. What are the conditions necessary to the preservation of sight?   We believe them to be a dominating mind and spirit in harmony with divine law. We believe them to be a knowledge and application of truth.

Physical blindness corresponds to spiritual blindness. Failure of physical sight eorresponds to an inability to continue to see truth. There is a universal failure to see truth as governing the reconstruction of our eyes. In youth we expect to have good sight; we fully believe it to be our right. In old age we have permitted our belief in truth to be argued away from us. We do not wonder that as a rule the spiritual blindness that surrounds us, added to our own, reflects itself in our failing sight. We only wonder that there are any exceptions to this rule.

A knowledge and application of truth can work all good. We know of cases in which persons have by spiritual science, preserved unfailing sight through advanced years, even against the obstacles of heredity and surrounding error. We know further of those, who after wearing eye-glasses for some years, have entirely discarded them, with constantly improving sight as a result.

The same arguments that apply to failing sight, apply also to any other failing faculty. When we become desirous above all things of seeing truth, we shall be able to preserve our sight. When we become desirous above all things of listening to truth, we shall preserve our hearing. When we become desirous above all things of remembering truth, we shall be able to preserve our memory, so as to remember also even less important matters, for the greater contains the lesser.

Our earnest desire for truth will bring us truth and irrespective of our age.

Our mind should be just as capable of a dominion over our body at ninety years of age as at twenty. There is no age at which a harmonious mind will feel obliged to quit work and leave itself a prey to dominating minds in error. Our material body not only should never be under the dominion of our mind in error, but it should always be under the active or passive dominion of our mind in truth, in order to protect it from the error that is all around us everywhere.

Instead of resignedly expecting that at a certain age, maladies and suffering or a general breakdown of health, must inevitably come upon us, let us so believe and trust in spiritual power, that at any age we shall be preserved, mind and body, in a sound, whole condition.

If we live in ignorance or defiance of divine law at twenty years of age, we give ourselves up to broken down health and disease at that age. If, on the other hand, we live in harmony with divine law at ninety years of age, we protect ourselves in health at that age.

Truth shall make us whole. Truth shall make us free. As we advance year by year in truth, let us feel that each new body, which year by year nature so wonderfully reconstructs for us, will be more sound and more whole than the last one, thus when our work is finished and we drop that body, we shall not enter into the spiritual realm through the unnatural and revolting gateway of disease.

How many cases we all know, of those who have dropped their bodies without disease! The grandfather perhaps, who has just finished a useful day's work, and with unimpaired digestion has enjoyed his evening meal, sits in his arm chair and quietly passes away. Of course the learned men of science, who have always turned their attention to matter instead of spirit, decide that there *muts* have been some hidden malady, to cause his departure. Even after an autopsy, in which no sign of an ailment is discovered, they still assert, that as there must be a cause for every effect, there has been some ailment, though beyond their detection. Their logic is sound, there is no effect without a cause, but studying only matter, the world of secondary cause, they are ignorant of that spiritual law by which one, when his work here is finished, and he is ripe and ready for a higher condition, can, in the midst of health, be attracted to another realm and depart thence as quietly and as naturally as the sound, ripe fruit falls from a tree.

There is much fine talk about departing this life in God's appointed time, but we think that very few depart this life in God's appointed time. They depart in their own time. Through sin or error they wander away from his laws and destroy their bodies. They may do so sinfully and consciously, as in the case of the intentional suicide, or they may do so ignorantly and unconsciously, by indulgence in destructive conditions of mind or weakly harboring disease. In the former case the deed perhaps is accomplished in an instant, while in the latter it may be the work of years; but in both they have equally destroyed their bodies, instead of waiting God's appointed time to go.

Of course whenever we go hence it is all for the best, and the only thing possible for us in our condition, the condition that renders it best and possible for us; but then if our condition is a wrong one, we should not be in that condition, and it is our work and not God's work that we are in that condition. We may not be *immediately* responsible for going before our time, but we are *immediately* responsible for the condition of mind that abruptly and unnaturally tore us from our material bodies. An intoxicated man may not be immediately responsible for a crime committed unconsciously; but he is immediately responsible for his condition of intoxication.

If we were in a high spiritual condition, instead of passing away in infancy and youth as is the case in our present stage of unfoldment, we would be quite likely to advance far beyond the three score and ten years allotted to man by a people even more material than ourselves.

In spite of this average regarding the age of

man, upon which we have allowed ourselves to fix, we constantly hear of those who pass far beyond this limit, which is an artificial and not a natural one. An encyclopœdia of human longevity, now being edited, contains sketches of twenty thousand persons aged one hundred years and more. All ages of the world and all races could furnish their quota to such a list, and yet we stubbornly hold fast to the idea that it is not natural to remain in this life beyond a certain fixed time.

We think that if we possessed the truth regarding human lougevity and were in a high condition, we would not only remain here an indefinite number of years, but we would remain with sound, whole bodies that would serve us well until our earthly work was accomplished. We would have no declining years, for instead of declining, or tending downward, we would, as we advanced, tend upward.

There is no such thing as old age for the soul that lives forever. There is no such thing as old age for the material body that is incessantly renewed.

Old age has no existence for us, except among the falsities of our erroneous thought.

CHAPTER IX.

Good and evil, like white and black, are the positive and negative conditions of life. God created good and man created evil.

Man by an abuse of his free will, turned his liberty into license, and evil was the result. Man by perverting his faculties, which were all intended for good, and employing them for his own selfish gratification, brought evil into the world.

If this wrong condition of things has been brought about and now exists, and we feel that we involuntarily enter into it, how are we to lift ourselves up out of it?

There is a wide-spread error regarding the sacrifice we are obliged to make in turning our attention to spiritual culture. Sacrifice implies the

(117)

giving up of some good for the sake of a higher good. But spiritual science, in harmony with the teaching of Jesus, enjoins upon us to give up only evil, and to hold fast to that which is good, and sacrifice does not enter into it at all. It points out to us the advantage, the economy of yielding up evil, in the form of sin, error and disease, and urges upon us to pursue only good in the form of truth and health, and also helps us to discriminate between the two.

To sweep away evil by simply denying its existence, is a summary and easy way of disposing of it; but unfortunately it fails to teach us what is evil and what is good in that heterogeneous collection of thought in the vast store-house of our confused and ignorant mind. While it is perfectly true that there is no evil as created and perpetuated by God, and that the only evil is our perversion of good, yet the mere assertion that all is good, may not help some, who are in the darkness of error, to see in just what way they *have* perverted good.

As Jesus seemed to realize when he enjoined upon us to overcome evil with good, we need to know what our evil is in order to know what to overcome. There are very few around us at the present day, as there were very few around him at that day, who are so unfolded that they can come up immediately out of their evil by the simple utterance of broad ultimate truth. In order to climb they need a ladder by which they can rise step by step.

After Jesus had healed the man who had been paralyzed for thirty-eight years, he seems to have thought that there was something more to do for him. The man needed to be warned regarding his future, and Jesus, after pronouncing him whole, told him to sin no more, lest a worse thing should come upon him. He turned the man's attention to the sin or error that had probably been the cause of his infirmity, so that if he had not realized it before, he was then made aware of the special evil that he had to overcome.

When Jesus told the woman of Samaria all things that ever she did, as she expressed it, he undoubtedly had in view some good for the woman. While he may have wished her to recognize his gift of intuition in order to arouse her faith in his teaching, he doubtless also desired to call her attention to the special error in whose bondage she was at that moment living, that she might more fully realize her guilt and strive to rid herself of the evil.

Sin and error, the primal causes of suffering and disease, arise from mistaking evil for good. We all, however depraved we may be, desire only good, and as soon as we learn what really is good, we are ready to drop evil and pursue that good. As soon as the scales are taken from our eyes, we will, like Saul of Tarsus, perceive the evil to be evil and turn our energies to good. Spiritual knowledge takes the scales from our eyes. It gives us the power to recognize evil and to bring it to the front, and vanquish it with good.

Spiritual science, instead of wresting from us all material good, as is sometimes thought, not only permits us that good in abundance, but teaches us the highest and most enjoyable use of it.

"Blessed are the meek, for they shall inherit the earth." Those who maintain that lowly and correct attitude of mind enjoined upon us by spiritual teaching, shall be the very ones who will attract and most enjoy material blessings.

Would there not be something revolting in the

intention of a Creator who could place us in the midst of material beauties and delights that were designed only as a snare to tempt us from the path of virtue? They were created for our enjoyment, and they are a snare to us only when we put them out of their place.

Material things are intended to minister to our pleasure just as long as we are in that stage of unfoldment in which we require their ministry. They were never intended to minister to our *highest* needs, and the very endeavor to make them the all in all, is a perversion of good—therefore an evil—that recoils upon us, who so misuse them, and deprives us of the highest enjoyment to be derived from them, which is the true inheriting of the earth.

A pure and spiritual minded lady, who was seeking and finding spiritual truth, took alarm when told that no true spiritual scientist could possibly make use of intoxicating liquors. She said, "What! Shall we be obliged to make ourselves unpleasant by refusing wine at a friend's table, and by having none upon our own? And will it not end in our being obliged to give up our dinners, and our carriages and our dress and our jewels and everything that seems to belong to our position in the world, and going into sackcloth and austerities?"

She had abundant wealth, and as far as she knew was using it nobly. She was, like the majority among the cultured of this age, in that stage of her development in which material beauties and refinements ministered to her needs, and it seemed to her like losing all joy out of her life to give them up. But spiritual science does not urge us to give up anything that is good, it only teaches us what things are good, and how to give material things their rightful place. So long as we desire material things there is no reason why we should not have them. That is what they are created for. They are simply a response to our capability of enjoying them. But we can perhaps *imagine* a degree of unfoldment somewhere in our future, in which we shall, instead of having felt constrained to give them up, simply have outgrown them. There is nothing sinful, there is nothing wrong in a fondness for material things, it is only an indication of our spiritual youth. The purest and best of children are fond of toys, which a few years later on have become nothing to them. A mother does not feel that at a certain age she must make her children give up their toys, for she knows that as soon as they are sufficiently unfolded and their minds are turned upon more important things, they will naturally drop them, and there will be no sacrifice in the case.

The wrong in our love of material things comes in when we cannot be happy without a certain desired quality or quantity of them. We should be in that harmonious condition of mind in which we enjoy them when they are rightfully ours, but are just as contented when deprived of them. We should love them for the power of doing good to others, which they bestow upon us, as well as for the pleasure we personally derive from them.

The use and enjoyment of jewels and beautiful fabrics that are the rightfully acquired possessions of any woman, are legitimately hers, and are intended for her. They bestow upon her a power for good to her fellow beings that she would not otherwise possess. Were she to replace her fine fabrics with sackcloth, and leave her elegant mansion for a hut, she could do nothing like the amount of good that is possible with her in that station, and with those surroundings that became hers in accordance with divine law working through her own peculiar development. If she has rightfully acquired wealth, it is because it is best for her to have it, but it is a gift for the right employment of which, she is wholly responsible. It is intended for her to make good use of it, to press the mammon of unrighteousness or material things into the service of good. A finely decorated house or an elegantly appointed entertainment may not in themselves, be particularly righteous, still they may serve to attract within one's sphere a vast number of those on whose lives one can exert an immense influence for good.

A greater responsibility accompanies wealth than is generally thought. If you have rightfully acquired wealth it is because a good use of it is possible with you, if you make the effort for it; it is because you are unfolded in just the way to make it your servant, instead of your slave, if you desire and strive to do so; it is because you have the ability to use it well if you can conquer self enough to give that ability full scope.

Giving pleasure to others by means of material things, while it is not the highest good, may be al the good of which you are at this moment capable. If so, it is *your* highest good. But when you come into more light you will learn that the highest use of material things is to make them serve as stepping stones to spiritual things, to make them bring under your influence vast numbers of the unspiritual with whom they will begin the work of enlightenment. These material things will be to you a necessary tool in your spiritual work. A candlestick, although it gives no light, is necessary in order that the candle may shed abroad its light. Material things help you to throw your light upon the masses.

In speaking of our legitimate right to material things, we would say that it is confined to such as are good. If we, by some perversion of good, have created evil in any material thing, we have no more right to use it than we had to create it.

The lady to whom we have referred, was alarmed lest spiritual science, into whose truths she desired to come, should oblige her to give up all enjoyable material things. She did not know where to draw the line. The line is to be drawn between those which are good and those which are evil. Spiritual science will not *oblige* you to give up anything; but it will so enlighten you that you will voluntarily let go your hold upon the evil. If you are once in the possession of spiritual knowledge, those material things which carry with them an element of evil, will be as offensive to you, as a pestilential atmosphere to healthy lungs.

A tree is known by its fruits. Any material thing that works an injury to yourself or others must carry with it an element of evil, and is therefore to be rejected. If man receives harm from something that he has invented, the evil may be traced back to the very invention and distribution of it. If he receives harm from something already created, he may rest assured that he has not yet discovered its rightful use, for God created nothing to injure man. There is no such intention in any of God's gifts to man. Spiritual teaching does not *force* us to give up even the most harmful things; but it gives us such light, such knowledge that we see evil as evil and voluntarily reject it.

To one who has once risen up into the pure atmosphere of truth, any unnatural stimulant or narcotic such as intoxicating liquors, tobacco, coffee, tea, opium or any other poisonous drug, would, by a perfectly natural law, be only offensive.

If you were in spiritual knowledge you would not have to be told that it is indulging in an evil to place wine upon your tables and thus lead your sons and daughters into intemperance, and place temptation before those who have already fallen into it. If you had once made the voice of the spirit your guide you would become so sensitive to contact with evil that a glass of wine set before you, would for you, contain the concentrated suffering of thousands of men in the blackness of misery, and the pangs and wails of hosts of agonized and starving women and children who have fallen a prey to this desolating evil. If you felt all this in an unnatural stimulant that was placed before you, would you need to be urged to abstain from it? Would there be any enjoyment in it for you? If you once realized that all the good that is *supposed* to reside in these artificial stimulants, can be found in a greater degree in something that contains no element of evil, in something that does not induce abnormal appetites and passions, could you call it hospitality to set them before your friends, or could you commit the *monstrous* folly of offering them to your children?

If you were in a full realization of God's intention regarding a use of the material body, could you admire in others, or much worse, could you adopt for yourself, such fashions in clothing as prevented a natural and upright gait, or such as crippled you in the use of your vital organs? Would you be able so to descend from your high estate as to adopt any harmful fashion in dress, the very invention of which is only a pandering to vanity or impurity.

One, who with a lofty motive in some good to her fellow beings, might find it necessary to be deprived of a normal use of her vital organs or of her members, might, nevertheless, by means of her trust in good, be sustained and protected in health. But you, who with no good motive thus misuse your material body are not, and *cannot be* protected in health. You may have no desire to propagate evil—and we think very few have such a desire—you may be only a weak slave to prevailing opinion, nevertheless you are sowing evil for yourself and for others to reap. You will suffer the bondage of your voluntary slavery in mental degradation and physical decline.

You who are rich in this world's goods, who occupy high places in the land, who are leaders of fashion or custom, what power for good you **throw** away when you so fail to realize truth that in **your** toilets, in your *salons*, and at your tables, it is evil instead of good that you sow broadcast.

How grand would be that woman who, in the midst of luxury, could yet in the dispensation of all her favors, so draw the line between good and evil, that her influence was all on the side of good.

You may think that your associations are so material that you cannot rise above them. But if you have materiality in your surroundings, it is because there is something in your own condition of mind that attracts for you that materiality. There is something in *you* that makes it possible for you. When you, yourself become more spiritual, your surroundings will modify themselves in harmony with your new condition.

When you come into a knowledge of truth, you will have all the courage necessary to enable you to teach rationality and purity by your own example, or rather you will be in that condition in which no courage is demanded, for you will not be able to avoid becoming, yourself, an impersonation of rationality and purity. The lily teaches its lesson of purity without effort, because it is, in itself, an impersonation of purity. You will *unconsciously* draw the line between good and evil. You will appropriate the good by mere elective affinity, and reject the evil by a natural repulsion.

Let us not indulge in the prevalent mistake of supposing that we are in so high a condition that we can partake of evil and not be harmed by it, or that we can partake of only a *little* evil and receive only a *little* harm. The injury we do ourselves and others by the *least* indulgence in evil is always incalculable; and the very fact that the little evil is tolerable to us, proves that our condition is *not* high.

It is the mingling of evil with good that works the greatest harm. It is the eating of the tree of the knowledge of good and *evil* that is death to spirituality. One who is living almost wholly in evil is not regarded with admiration by even those as benighted as himself. His example is not likely to do as much harm as that of one who mingles a little evil with much good. Where the good predominates, the whole is taken for good, by the ignorant.

Let those especially who are in a conspicuous

position in the world, not indulge in the *little* evil, the harmful effects of which tell, not only on themselves morally, mentally and physically, but, tell in the same way upon incalculable numbers of envious and admiring beholders. Let such persons make it their highest pleasure *manifestly* to draw the line between good and evil.

There is a class of men, who are so blinded by their inordinate desires, that they are not able to discriminate between good and evil. It is so their habit to yield to the dictates of their lower nature that their higher nature has no chance to assert itself. Their divine soul has become so incrusted over with materiality that it seems to have departed from them altogether. They have, for the time being, lost their spiritual discrimination. Their whole course is one grand mistaking of evil for good. If you were to assert to them that all is good, they would quite agree with you and they would think that they possessed that good in the gratification of self. Nothing could awaken them from their slumber in falsities, except that which comes to them sooner or later, that which their own condition brings upon them, viz., intense suffering.

There is another class of men, who have partly
awakened from their dangerous sleep. To these
we would speak, for they can comprehend us.
They desire to drop evil out of their lives, and endeavor to do so just as fast as they learn what it
is. They are ignorant and bewildered and confound good and evil. To such we would say, let
this be your test; the indulgence in any practice
or the commission of any act that works harm to
yourself or others, must be an evil. Evil no more
comes out of good than corrupt fruit comes from
a sound tree.

If with specious arguments you are endeavoring
to persuade yourself that a moderate use of intoxicating drink or tobacco, or any other undue stimulant, is a good instead of an evil, and, in looking
around you searchingly, you find them working an
injury to thousands, you will be convinced of your
mistake, and you will wish to leave them entirely
out of your life.

By careful thought, even on the material plane,
you will learn that whatever gives pleasure or
comfort by inducing unnatural action or unnatural
torpidity in the mind—which reflects itself in a
similar bodily condition—works against the laws

of our being and is therefore an evil. A temporary comfort gained in this unnatural and therefore
illegitimate way ends in discomfort. Temporary
pleasure so gained, ends in misery. In the light
of mere worldly policy it is only adding mortgage
after mortgage to our estate it is only borrowing
to repay with compound interest. Real, true good
is not exciting in its nature, neither is it deaden-
ing It is always quickening; but in imparting
life, it imparts peace and tranquility. The greatest
powers do not work with an external commotion.
When we employ anything but good to incite us
to action, and push on our life thinking that we
are thereby increasing our vitality, we are in reality only decreasing it.

If we give our best mental powers to the subject, we learn that true spirituality, which is allow-
ing our higher nature to take the lead—and it is
always ready and willing to do so if we do not obtrude our lower nature in its way—is the most
rational, the most practical, and the most economical plan of life of which the human mind is able to
conceive.

If it is practical and economical for the farmer
to uproot the tares from his field, lest they choke

the grain, it is practical and economical for us to uproot the evil from our lives that the good may have a chance to grow. And as it is the farmer's first work to learn what, among his sprouting germs *are* tares and what are grain, so it should be our first work to learn to distinguish between evil and good. The wise farmer does not wait to let his tares come to maturity in order to learn that they are useless or noxious, but as soon as he knows that their tendency is harmful rather than beneficial, instead of saying that he likes them and that a few of them will not do much harm, the mere knowledge that they *are* tares and *not* grain, is sufficient to induce him to uproot them every one.

Why can not we work our spiritual fields in accordance with the same practical and economical plan?

If business men proceeded strictly in accordance with this plan, they would not ask themselves if a certain transaction or a certain course was legal, or customary among business men, but they would only endeavor to learn whether it was beneficial or harmful in its tendency.

A man who has an inordinate desire for wealth

or fame, is not in a condition to discriminate between good and evil. The gratification of his ruling desire is his only good. He sees everything
through the distorting medium of his self-hood.
If he would see clearly, let him remove self from
before his vision. Let him think of the good he
can bestow upon others, instead of dwelling upon
the good he can grasp for himself. A sudden
turning of his mind away from evil and towards
good, may, and doubtless will, for the moment, be
as painful to him as the amputation of a decaying
member of his body, but it will, in the end, prove
equally salutary and economical.

It is such a mistake to suppose that the good of
this life is a contradiction of the good of our so-
called future life. There is but one good, and it
applies to our whole continued existence. All
good is of the spirit, and only reflects itself in
matter. I we would have our life in matter, the
perfection of a material life we must place it under the rule of the spirit which alone has access to
good. Our conscience must be our gate-keeper.
If we put out the eyes of our keeper or shut him
up in the dark until he has lost his sight, he cannot guard our premises, and they will be exposed

to the inroad of encroaching evil-doers. We all have a conscience, but if we blind it by the indulgence of selfish desires, it will not distinguish between good and evil, and we shall be invaded by foes that have been mistaken for friends.

How shall we cultivate our conscience, on whose faithful discharge of duty depends both our spiritual and material prosperity? How shall we teach it to become so discriminating, so on the alert that no evil, however well it may counterfeit good, can come in upon us?

We can do so only by endeavoring to benefit others equally with ourselves, by studying the interest and welfare of others as we study our own interest and welfare. It is only by living up to the highest light we now possess that we can gain more light, and be enabled to distinguish evil from good.

CHAPTER X.

———

———

A true science of life must apply to all conscious, living beings. A knowledge of the right mode of thinking and living, applies to all beings who consciously think and live, and it applies to them at all times, and in all places, and under all circumstances. It applies to man, woman and child. It applies to infancy, youth and old age. Unlike any other science or knowledge, it has a bearing upon everything in life, upon every conceivable occupation, and upon our pleasures as well as upon our duties.

There is, among the ignorant, a mistaken idea that only those who lack bodily health need trouble themselves about this kind of knowledge. Of course those who lack bodily health, thereby *reveal*

(137)

their lack of this kind of knowledge. But those, who at this moment seem to have bodily health, need the teachings of this science to enable them to maintain that health, to protect them against encroaching outside influences, and, above all, to enable them to correct their own errors to whose harmful effects they will otherwise sooner or later succumb.

Then giving us bodily health is the work the science accomplishes on the very lowest plane. It is merely the indication that the higher and more interior work has already been effected. The science aims at the higher work and not at the lower.   It aims at the greater work and not at the lesser, and in effecting the greater, which includes the lesser, it must accomplish both.   When we build a fire in our room it is to keep us warm, and not to raise the mercury in our thermometer, but it accomplishes both.

A true spiritual science directs its efforts to spiritual regeneration, and from the spiritual rebirth naturally results the physical rebirth or reformation.   So long as we fall short of perfection we need a spiritual rebirth.   We all, without exception, need more spiritual knowledge.   We need not only to gain that knowledge, but to utilize it, and utilizing it is bringing it to bear upon every feeling, thought and act of our whole life.

There are certain classes and societies of people, who are accomplishing great and noble work in some one line of thought, who, nevertheless, in their ignorance of its true nature, antagonize spiritual science, instead of seeking it as a means whereby their power for good in any direction whatever might be increased a thousand fold.

Any knowledge as well as any good intention in the way of philanthropy and reform is included in spiritual science, although it may be an infinitesimal part of that science. If your design is evil you may consistently oppose spiritual science, for it will not aid you. It will not help you to take advantage of your neighbor, or to grasp what does not honestly belong to you, or in any way to live for your own selfish gratification. But if your design is good, so far as it *is* good, just so far you have already accepted the truths of this science, whether you are conscious of the fact or not. But a consciousness of the fact and an endeavor to increase your knowledge, would increase your power for the special good you have in view.

If, for example, you believe that the tests of spiritualism as offered in material phenomena help the world to a knowledge of the immortality of the soul, and of true spiritual life, why do you not also see that spiritual science can help you in this very work? Why can you not perceive that it includes the teaching you are endeavoring to give?

If you are working in the cause of temperance, why can you not understand, that the only radical and reliable reform in that respect, must be in the spirit and mind? It must result from that very temperance and moderation of thought, that very spiritual control of the passions and appetites to which we are so lovingly assisted by spiritual science. If you were working to increase intemperance you would be consistent in opposing this science, for it could not espouse your cause. But as you really desire good, your good, whatever it may be, is included in the science.

If you are working to educate the young, and take an exaggerated view of the importance of the intellect, remember that the highest kind of educating or unfolding is that to which we are led by spiritual science. It is a kind that develops our highest powers, that unfolds both the intellect and the spirit, and by the very development of the spirit, renders even the intellect, which yon over-rate, still more capable than it could become under your one-sided unfoldment. If you were working to keep the world in ignorance, you would natural-ly antagonize this science, for it would be only a constant rebuke to you. But as you are seriously working for the spread of knowledge, you can do nothing wiser than to throw youself heart and soul into this science which will enable you to sow broadcast the very highest kind of knowledge.

If your efforts are directed to the uplifting of woman, to the placing of woman on an equality with man, you, above all others, should be attract-ed to this divine science that recognizes no sex. If you think that suffrage is a means whereby woman will be protected and elevated, and man also refined and improved, you can do nothing more strengthening to your cause than to turn your attention to this science, which gives every human being a voice in the governing of his own life, as well as an influential voice in the govern-ing of all other lives into whose sphere he enters. It is a science whose whole tendency is to refine

and elevate, and it offers the only perfect freedom
to all, irrespective of race, color, creed or sex.
For you to antagonize this science, would appear
as if you were desirous to enslave woman and en-
courage despotism in man.   If you sincerely and
wisely have at heart the cause you pretend to espouse, instead of opposing a willing ally, you will
desire to join your forces to all that are working
in the same direction as yourself.   You will see
that the most effective way to increase your own
power is to recognize and adopt this science in
which resides the very essence of power.   You
will perceive that the divine scheme of life includes
the whole of your aim, as one of its many parts;
and that all its parts have a bearing upon one another, and that each partakes of the quality of the
whole, as a drop of the ocean partakes of the
quality of the great body of water as a whole.
You will learn that the science of life includes the
whole of life, and therefore you cannot fail to find
therein the small part to which you have directed
your energies.

Our one little aim, however good it may be, is
only one atom of a great whole.   Let us, as far as
possible, come into a knowledge of that great

whole. Let us learn that each good, however dis-
tinct and supreme it may seem to us who have
narrowed down our vision to its outlines, is yet
only one of many kinds of good, and that all kinds
are so related that they cannot reach full stature
independent of one another. A knowledge of
more good will give us a broader view of our own
little good, and increase our usefulness in its pur-
suance.

Let us not fear that our little code of morality,
or our little religious sect will be overthrown if we
extend our knowledge of even *possible* truth. That
which is not truth will not stand, though it may
lead us nearer truth than we were before, as a dark
passage way may conduct us to a light room. Do
not let us fear to use our powers of discrimination
in order to test any pretended truth, for if it is not
worthy it will not bear a test.

Let us first be perfectly sure that we have come
into possession of this science of our interior life,
then let us daily and hourly apply it, and we *must*
apply it if we possess it. We shall then realize its
truths to the extent of our ability to realize them,
to the degree to which we are unfolded intellectually and spiritually. We shall be able to see that

it has a fellowship with all that is noble and aspiring in even the most degraded of races and of individuals. We shall realize its universality and our love and our charity will become so universal that they will include the whole of mankind. We shall realize its immensity and we also shall become great. In order to do great work we must become great ourselves.

Although this science most certainly teaches us how to live in order to keep our material body in good condition, yet the teaching is not cast aside with the body. It is eternal truth. In gaining it we are laying up treasures where neither moth nor rust doth corrupt nor thieves break through and steal.

In this material age much thought is devoted to making good investments of the money that comes into our possession. We feel so satisfied and secure when we have put it into something that we consider *sure*, something that cannot fail to bring us a due return, when the truth is that the very safest material investment one can possibly make is about as secure as a bubble in a tempest. There is nothing fixed, nothing unchangeable in the world of matter.

If, on the contrary, we invest all our best faculties, all our highest powers in the divine enterprise that works only in the realm of spirit, we may well have a feeling of 'security. We have placed our wealth where there is no such thing as change. We have invested it where its interest is momentarily compounded, and where it is protected by laws that are as fixed and eternal as the Infinite Being who has charge of it, and in whom there is no variableness, neither shadow of turning.

A knowledge of our spiritual nature and of the laws by which it is governed, once gained, can never depart from us. A truth that is once ours has become a part of our divine being, and can no more be detached from us than any other part of our immortal self.

Let us strive to gain these everlasting riches, then shall we be rich indeed and in truth.

Practical Metaphysics,

By M. J. BARNETT.

A CONCISE STATEMENT OF THE PRIN-CIPLES OF METAPHYSICS.

This work, on account of its clearness of expression and its practical value, has been adopted as a text book in teaching the science of metaphysics or spiritual healing. It is recommended also to the sick and afflicted who have no access to teaching other than through literature, and who find it, what it professes to be, the divine science of healing made practical.

RETAIL PRICE, - $1.00.

Sent by Mail, Post-paid, on Receipt of Retail Price.

HEALTH FOR TEACHERS.

BY

M. J. BARNETT,

AUTHOR OF

PRACTICAL METAPHYSICS, &c.

BOSTON:
H. H. CARTER & KARRICK,
3 BEACON STREET,
1888.

JUSTICE

A HEALING POWER.

BY

M. J. BARNETT,

AUTHOR OF

"Practical Metaphysics, or The True Method of
Healing," "Health for Teachers," etc.

BOSTON:
.... H. CARTER & KARRICK,
3 BEACON STREET,
1888.

MENTAL MEDICINE:

A THEORETICAL AND PRACTICAL TREATISE ON MEDICAL PSYCHOLOGY.

By W. F. EVANS.

This book contains a full exposition of the nature and laws of Magnetism, and its application to the cure of disease.

Extra Cloth. 216 pp. Retail Price, $1.25.

☞ *Sent by mail, postpaid, on receipt of retail price.*

SOUL AND BODY;
OR,
THE SPIRITUAL SCIENCE OF HEALTH AND DISEASE.

By W. F. EVANS.

Extra Cloth. 147 pp. Retail Price, $1.00.

☞ *Sent by mail, postpaid, on receipt of retail price.*

All the above works of Dr. Evans are on the relation of Mind and Body, and the cure of disease in ourselves and others by the mental method, and are the only publications on the subject that commend themselves to men of science and to thinking people everywhere.

THE WORKS

OF

DR. W. F. EVANS.

THE INFLUENCE OF THE MIND ON THE BODY IN HEALTH OR DISEASE, AND THE MENTAL METHOD OF TREATMENT

———

" On earth, there is nothing great but Man:
In Man there is nothing great but Mind."

———

PUBLISHED BY

H. H. CARTER & KARRICK,

3 Beacon Street, Boston.

1886.

THE DIVINE LAW OF CURE.

By W. F. EVANS.

A Standard Work on the Philosophy and Practice of the Mind Cure, a Reliable Text-Book in all the Schools of Mental Healing.

No work has ever been written of more practical value to physicians of all schools. The book is the result of the extensive learning and research of the author and exhibits a familiarity with the literature of the subject. It is profoundly religious without being offensively dogmatic. It has been received with universal favor by all who are seeking light on the subject on which it treats — the cure of disease in ourselves and others by mental and spiritual agencies.

Extra Cloth. 302 pp. Retail price, $1.50.

☞ *Sent by mail, postpaid, on receipt of retail price.*

The Nature and Power of Faith;

OR,

ELEMENTARY LESSONS IN CHRISTIAN PHILOSOPHY AND TRANSCENDENTAL MEDICINE.

By W. F. EVANS,

Author of "Divine Law of Cure," etc.

This work is a complete exposition of the principles underlying the system of mental healing. It contains a full course of instruction in the philosophy and practice of the Mind Cure. It is the most complete treatise on Christian Theosophy, in its application to the cure of both soul and body that was ever published. It has elevated the subject into the dignity of a fixed spiritual science.

Extra Cloth. 225 pp. Retail Price, $1.50.

☞ *Sent by mail, postpaid, on receipt of retail price.*

www.ingramcontent.com/pod-product-compliance
Lightning Source LLC
Chambersburg PA
CBHW021813190326
41518CB00007B/580